Adaptation to Climate Change in Developing Countries

Adaptation to Climate Change in Developing Countries

Edited by Daisy Mathews

SYRAWOOD
PUBLISHING HOUSE
New York

Published by Syrawood Publishing House,
750 Third Avenue, 9th Floor,
New York, NY 10017, USA
www.syrawoodpublishinghouse.com

Adaptation to Climate Change in Developing Countries
Edited by Daisy Mathews

International Standard Book Number: 978-1-64740-341-6 (Hardback)

Cataloging-in-publication Data

Adaptation to climate change in Developing countries / edited by Daisy Mathews.
p. cm.
Includes bibliographical references and index.
ISBN 978-1-64740-341-6
1. Climatic changes--Developing countries. 2. Climatic changes--Government policy--Developing countries. 3. Climate change mitigation--Developing countries. 4. Developing countries--Environmental conditions. 5. Adaptation (Biology)--Developing countries. I. Mathews, Daisy.
QC903.2.D44 A33 2023
363.738 74--dc23

Table of Contents

PREFACE

This book has been an outcome of determined endeavour from a group of educationists in the field. The primary objective was to involve a broad spectrum of professionals from diverse cultural background involved in the field for developing new researches. The book not only targets students but also scholars pursuing higher research for further enhancement of the theoretical and practical applications of the subject.

Climate change adaptation refers to the process of making adjustments in the anticipated or existing effects of climate change. It is one of the strategies for responding to climate change in conjunction with mitigation. Adaptation actions can be transformative or incremental. The need for adaptation differs in different areas depending on the vulnerability and sensitivity to environmental influences. Adaptation actions are classified into three types including structural and physical adaptation, institutional adaptation, and social adaptation. It is particularly crucial in developing countries as they are the most sensitive to climate change and are facing the brunt of its consequences. This book contains some path-breaking studies on adaptation to climate change in developing countries. It picks up individual strategies and explains their need and contribution for mitigating climate change. This book will serve as a reference to a broad spectrum of readers.

It was an honour to edit such a profound book and also a challenging task to compile and examine all the relevant data for accuracy and originality. I wish to acknowledge the efforts of the contributors for submitting such brilliant and diverse chapters in the field and for endlessly working for the completion of the book. Last, but not the least; I thank my family for being a constant source of support in all my research endeavours.

Editor

1

Climate Change Adaptation: Implications for Food Security and Nutrition

Caroline Fadeke Ajilogba and Sue Walker

Contents

C. F. Ajilogba
Division of Agrometeorology, Agricultural Research Council – Soil, Climate and Water, Pretoria, South Africa
e-mail: ajilogbac@arc.agric.za

S. Walker (✉)
Division of Agrometeorology, Agricultural Research Council – Soil, Climate and Water, Pretoria, South Africa

Department of Soil, Crop and Climate Sciences, University of the Free State, Bloemfontein, South Africa
e-mail: walkers@arc.agric.za

Abstract

It is apparent that climate change affects every facet of life as no living organism lives outside of the universal environment (air, water, land), and all of these are affected by one or more climate or weather conditions. Climate affects agriculture and vice versa as they are deeply interconnected. The effect of climate change on agriculture goes a long way to increase or decrease food security and invariably food nutrition through its impacts on agricultural lands. This is because the more food insecurity there is, the more issues of undernutrition are observed.

This chapter studies the different ways climate and climate change affect crop production through the different sections of agriculture in terms of plant diseases and biocontrol, food production, livestock rearing, fish production, forestry, and microbial diversity. It goes further to look at the different ways nations and communities are adapting to climate change to mitigate the challenges of food insecurity and nutrition. Finally, some of the solutions that can be pilot tested at the community level, which can later be cascaded to national and regional levels, are also emphasized. Other recommendations that can become a research focus to forestall this threat are also highlighted and would be important in policy development.

Keywords

Climate change · Environment · Food availability · Food production · Food security · Nutrition · Soil · Weather

Introduction

Food security has been affected by a change in climate and weather conditions globally. Recently it is also obvious that food nutrition security needs to be addressed. Agriculture is important for food security in two ways: it produces the food people eat and (perhaps even more important) it provides the primary source of livelihood for 36% of the world's total workforce (ILO 2007). The Food and Agriculture Organization (FAO) of the United Nations (UN) defined food security at the World Food Summit in 1996 as "when all people, at all times have physical and economic access to sufficient, safe and nutritious food to meet their dietary needs and food preferences for an active and healthy life" (Food and Agriculture Organization 1996). Food security transcends food availability (production, distribution, and exchange) but also include how food is accessed (affordability, allocation and preference), utilized (food safety, nutrition, and social value) and also how it is distributed in the supply chain (food stability) (Ziervogel and Frayne 2011). This also means that it is not just about food but how available food is. And even when it is available, communities and individuals might be food insecure if they are not able to access food either because of monetary incapacitation or because of lack of knowledge of where to acquire the food, based on their livelihoods and socioeconomic level. It could also be as a result of how safe the food is especially in terms of processes followed during food preparation (Ziervogel and Ericksen 2010).

Agriculture, forestry, and fisheries have all been affected by climate change, and they have contributed to the increase in greenhouse gas emissions, thus affecting climate change negatively. The solution to climate change also partly lies within this sector, if agriculture is able to mitigate climate change through reduction of greenhouse gas emission during sustainable agricultural practices (Al et al. 2008).

Adaptation to climate change can constitute a range of interventions across all levels from individuals to communities to regional and national level. As observed by Bierbaum et al. (2013), communities can initiate their climate adaptation efforts in numerous ways. For example, in the United States, some diverse strategies employed by different communities to combat the issue of climate change include:

- Focusing on internal operations
- Beginning with a sector that is particularly vulnerable to climate change
- Integrating climate change concerns into existing planning approaches
- Conducting a community-wide assessment

The results from these experiments with adaptation strategies demonstrate that there is no "one size fits all" adaptation process or adaptation action (Bierbaum and Stults 2013). This is quite important to note in adaptation plans; however, the information required and the approached to adaptation can be similar across regions and sectors. Therefore, strategies that encourage data and best practices sharing, practical approaches, collaborative, and integrative processes will be important in developing adaptation plans in different countries.

Climate Change and Reality

Climate change occurs when changes in the Earth's climate system result in new weather patterns that last for at least a few decades and maybe for millions of years. Effects of climate change are seen in the frequency and severity of extreme weather events including droughts, floods, new rainfall patterns, and heatwaves. Increases in temperature have led to an escalation in the distribution of weeds, pests, and diseases some of which are a threat to marine life, plants, animals, and humans. In the same vein, an increase in extreme weather events has led to soil erosion (by water and wind) and loss of arable land for forestry and agriculture.

Impact of climate change on food production cannot be overemphasized; it is enormous and felt more acutely in developing and the least developed nations (Fig. 1). Flooding and droughts have been observed to ruin crops and pastures already being cultivated on the fields. Losses due to floods and droughts between

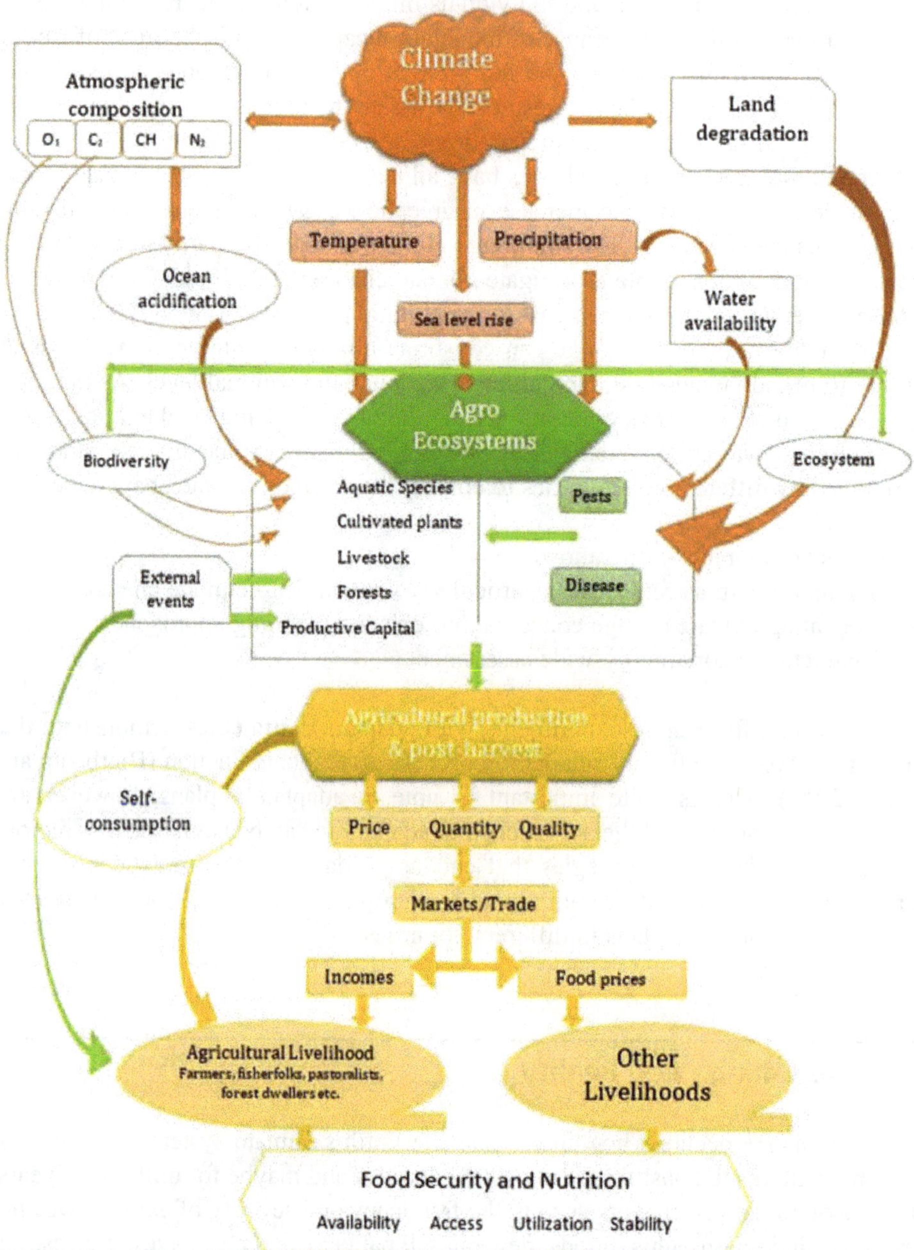

Fig. 1 The interrelatedness between climate change, food security and nutrition showing the impact of climate change on food security. (Adapted from FAO 2016)

2004 and 2014 have been estimated to be over 100 billion USD globally (Food and Agriculture Organization 2015; Brida et al. 2013).

Agriculture and fisheries are extremely climate dependent. Temperature and carbon dioxide (CO_2) increases in some areas can increase some agricultural productivity. Nevertheless, the higher demand for nutrients, soil moisture, water availability, and other requirements must also be met in order to realize these benefits. Changes in drought and flood frequency as well as intensity can present severe threats for farmers and ranchers and therefore threaten food safety (Ziska et al. 2016). In addition, warmer water temperatures are likely to trigger a change in the habitat ranges of many species of fish and shellfish, which may threaten ecosystems. Furthermore, climate change could make it harder to produce food and fish and raise livestock in the same way and in the same places as in the past (Chen 2015).

Climate Change Adaptation and Mitigation

Adaptation to climate change is a way of responding to global climate change (also called "climatic change" or "anthropogenic climate change"). According to The Intergovernmental Panel on Climate Change (IPCC), adaptation is sequence of modifications to actual or expected climate and the impacts of the modification. Adaptation is aimed at mitigating or preventing harm or manipulating beneficial possibilities within human systems. Sometimes, the contributions of human beings enhance adaptations to these modifications and its impacts (IPCC 2014). This requires intentional efforts in the human environmental system to adapt behavioral patterns, lifestyles, and perceptions that can lower the risk to human lives and livelihoods, while mitigation involves coordinated actions to reduce long-term emissions of greenhouse gases (Al et al. 2008). In contrast, climate change mitigation means reducing greenhouse gas emissions and short-term sequestration and processing of coal and even more significantly by making infrastructural decisions that will reduce risk by reducing long-term emissions. While the entire food system is a producer of greenhouse gas emissions, still the most important component is primary production. Incentives are required to encourage producers, agribusinesses, and managers of biodiversity to follow good practices to mitigate climate change.

Food Production and Climate Change

Climate change will affect food production and welfare of food animals (i.e., livestock, poultry, and fish) in several different ways. Consider the production of livestock that will probably be restricted by climate change as consumption of water by livestock is expected to increase by a factor of three. The need for arable land for pastures and feed production will increase because of a 70% increase in demand for agricultural products. Food safety issues also become important because around one-third of the world's cereal harvest is used for animal feed (Rojas-Downing et al. 2017). Furthermore, 14.5% of global greenhouse gas (GHG) is produced by the livestock industry. The livestock sector will therefore be a key player in reducing GHG emissions and enhancing global food security in the future (Rojas-Downing et al. 2017; Gerber et al. 2013).

Climate change and competition for land together with water and food security are likely to adversely affect livestock production at a time when it is most wanted (Thornton 2010). The output efficiency of feed conversion ratios for food, dairy, and egg production could be felt as a result of losses due to climate change. Increasing temperatures can also lead to death of food animals, as was noted by the deaths of more than 30,000 dairy cows in California and also in reduction of the overall production of milk in 2006 (Pitesky et al. 2014). Temperature affects most of the critical factors for livestock production, such as water availability, weight gain (feed conversion ratio), reproduction (semen quality, cow calf efficiency), and general health. A combination of increases in temperature, CO_2, and variability in precipitation influences the quantity, quality, and availability of forage and grain feed crops. Livestock diseases are affected primarily by an increase in the variation in temperature and precipitation (Thornton 2010).

Changes in land use, food production, animal production, manure, storage and transportation, and livestock all influence local and global weather patterns. Production of feed and manure releases CO_2, nitrous oxide (N_2O), and methane (CH_4), thereby influencing climate change. Therefore, livestock production systems increase emissions of CH_4. The production and transportation of animal products and improvements in land use contribute to increased CO_2 emissions. Research on the impact of climate change on livestock production is still minimal and need renewed attention (IPCC 2014).

Insects are ectotherms and are highly sensitive to their surrounding environmental conditions particularly the temperature; therefore temperature is a major factor affecting their reproduction (Kambrekar et al. 2015). This means increases in temperature for both edible and disease-causing insects could increase their reproductive cycle up to five cycles in a single year, which could affect farm food production negatively in terms of crop stress and yield loss. For beneficial insects, this increase in temperature will lead to increase in their metabolism (Kambrekar et al. 2015). Perhaps this provides an opportunity to develop new food systems based on insect production, either as feed for livestock and fish or directly for human consumption.

The possible effects of changing climate on insects could result in outbreaks, migration, change in biodiversity, species extinction, change in host shift, and emergence of new pests or biotypes. Due to the climate change, there is an increase in insect pest population, more outbreaks of insects, increased number of generations, as well as development of resistant biotypes (Moore and Allard 2008).

Fish plays an important role as a source of nutrients and micronutrients for human consumption, and fishing is a crucial food-producing sector and a source of livelihood for coastal farmers. Both fisheries and aquaculture make a significant contribution to food security and the livelihoods of millions of people as job creators, nutritious food providers, revenue generators, and economic growth through harvesting, processing, and marketing fish. Especially susceptible to climate change are small-scale fisherfolk and small-scale aquaculture. Their vulnerability is the function of both their location and their circumstances of vulnerability and disadvantage. Fishing and fishing communities are exposed to extreme climatic conditions and natural hazards, such as hurricanes, cyclones, sea level rise, ocean acidification, flooding, and coastal erosion, as they are mostly situated on the coastline. Millions of people living in the lowlands of coastal regions and the floodplains cannot escape regular flooding. Climate change affects living conditions as well as physical well-being of humans across many agro-forest ecosystems of the world (FAO 2016).

Forests are a climate-stabilizing force. They control habitats, preserve biodiversity, play an integral role in the carbon cycle, improve living conditions, and provide goods and services that can boost sustainable growth (Bonan 2008). A forest's position in climate change is that it both causes and mitigates greenhouse emissions; therefore, there are both pros and cons. Impacts of climate change on forests result from variations in atmospheric CO_2 concentration, and shifts in temperature and precipitation regimes, as well as an increase in climate variability reflected by increasing frequency and severity of extreme events. Change in climate will affect tree photosynthesis, rates of growth, leaf phenology, seed production, and cycling of nutrients (van der Meer et al. 2002). In some regions, forest form a vital resource for the community. For example, tropical forests cover about one third of the Kafa highland region of Ethiopia where a multiple tenure system exists for the utilization of the forest resources (Zewdie 2003). These include sharecropping and allocation of trees to sons when they come of age. However, the beekeeping and wood processing skills also need to pass from one generation to the next to maintain sustainable livelihoods (Zewdie 2003). Therefore, climate change effects on such forests will have a major impact on the livelihoods of the people.

The land management sector, the second largest driver of greenhouse gas emissions after the energy sector, accounts for around 25% of global emissions. Around half of these come from deforestation and habitat destruction (5 10 Gt CO_2e per year). Planting of forests and commercial plantations are among the most important strategies to address climate change effects. It is estimated that 2.6 billion tons of carbon dioxide, which is one-third of CO_2 from burning fossil fuels, is absorbed annually by forests. Projections show that there are nearly two billion square kilometers of worldwide degraded land – an area the size of South America – that can open up opportunities for restoration. Therefore, forest growth and preservation is an important solution to address the impact of climate change.

There are four major strategies available to mitigate carbon emissions through forestry activities:

- By reforestation (Bierbaum et al. 2013)
- Increasing the carbon content of natural forests on both stand and landscape scales
- Expansion of the use of forest products that replace CO_2 emissions from fossil fuels sustainably
- Reducing deforestation and degradation emissions

Climate also adversely affects the production of food crops that are highly sensitive to climate parameters. Long-term trends in average rainfall and temperature, interannual variability in climate, shocks during specific phenological stages, and extreme weather events affect food production (IPCC 2012). Some crops are more tolerant to certain stress types than others, and different stress types affect each crop species in different ways at each phenological development stage (Simpson 2017). Even though, crop yields are expected to decline under climatic conditions in the future, crop yields are already being affected as indicated by recent research (Ray et al. 2019).

As climate changes, crop production strategies must change also. There will always be some uncertainty associated with modeling the complex relationships between agricultural yields and future climate scenarios.

Climate Change Impact on Food Availability

Food availability is determined by the physical quantities of food that are produced, stored, processed, distributed, and exchanged (Al et al. 2008). Food production is affected by climate especially in regions where biochemical processes are being influenced by climate change. Some of these influences could be positive, negative, or even neutral (Allen Jr 1991). Location is a determinant of how the effect of climate change will be experienced in different regions (Leff et al. 2004). In temperate regions where elevated levels of CO_2 will increase plant growth and development, greenhouse fertilization effect is beneficial so crop yields will also increase. Such increase is observed to be as high as 0–10% increase in crops with very high photosynthetic efficiency, while those crops with lower efficiency have higher increase from between 10% and 25%. In such scenarios, the assumption is that the level of CO_2 in the atmosphere will increase to 550 parts per million (Tubiello et al. 2007; IPCC 2007). In some cases, increase in temperature between 1 °C and 3 °C is beneficial to plants in temperate regions; however, in tropical and seasonally dry regions, the impact is negative on crops especially cereals that are already growing near to their optimal temperature regime. Nevertheless, the impact is negative on all crops when the increase in mean temperature is more than 3 °C (IPCC 2007).

Such temperature increase will also impact livestock production as the growth of feed crops is also compromised and become an indirect effect on livestock production (Porter and Semenov 2005). Direct effects are related to exposure to solar radiation that then affects the exchange of heat between the animal and the environment (Rust and Rust 2013). This is critical, as many livestock breeds are not tolerant of high temperatures (Rust and Rust 2013). Forage production from rangelands in Africa is sensitive to changes in climate and will have a substantial impact on the livelihoods of more than 180 million people who raise livestock on rangelands (Boone et al. 2018). Climate effects on forage quality that depends on nutrient composition and affects digestibility, partitioning of metabolized products in the digestive tract, and intake of forage by animals will strongly affect animal performance.

Some drastic changes in the farming system sometimes have to be introduced, such as a change from sheep and cattle to goats, as they are the only domesticated animals with a versatile ability to adapt to a changing climate more readily than any other ruminants. Generally, goats are less affected by the harsh hot dry climates compared with other ruminants. Some of their ability to adapt is due to their habit of browsing together with the anatomical advantage of the upper lips. Therefore, goats can thrive well with limited feedstuffs, especially in arid and semiarid regions (Pragna et al. 2018). Another advantage is that during feed scarcity, goats can reduce their metabolic processes to conserve energy resource (Yadav et al. 2013). Therefore, they provide a continued supply of meat despite increases in temperature and frequencies of drought.

Climate Change Impact on Food Accessibility

"Climate change is becoming something we can taste" (Little 2019). According to Little (2019), this is the age of diversity and accessibility of food. Food accessibility is a function of being able to maintain and manage what one is entitled to, which is defined as the collection of assets (including legal, political, economic, and social) needed by a person to gain access to food (Sen 1989, cited in FAO 2003, Sage 2014).

Food accessibility refers to ability to own food, distribute it, and make food choices which allows people to turn their hunger effectively into demand (Sage 2014). Since food accessibility focuses on the power of households/individuals to be able to spend money, and the group dynamics of access to food, poverty, and vulnerability are key functions of it (Ziervogel and Frayne 2011).

Households have been observed not to grow their foods independently but either buy, trade, or exchange them (Du Toit and Ziervogel 2004). The ability to acquire food by households can be impacted by climate as it influences either the job opportunities or the availability of such foods may influence non-availability of some foods leading to increase in price, which could also make such foods unaffordable, thereby impacting on individual's nutrition and health. When there is an increase in price, low-income earning families are at the receiving end, so when they cannot purchase such foods, they either eat less preferred foods or reduce quantities of food eaten, making food prices very important to food security (Pretty et al. 2005).

Climate Change Impact on Food Utilization

Food utilization refers to how food is used and how individuals can obtain vital nutrients from the food they eat. This includes the nutritional benefits of the food, the nature and method of processing the food, and the way the society views each food that also determines when if and when such foods should be eaten. Also included are the knowledge of the quality and health of the food, which determine if there would be nutrient loss or the susceptibility of the food-to-food-borne diseases if below standard (FAO 2008).

When there is insecurity of food, dietary diversity is reduced, and this could lead to non-availability of nutritious food that form basis of balanced diet of the populace. This scarcity could be as a result of water scarcity, drought, or labor from climate change and variability. This impact of climate change on food utilization is indirect because its effect is ultimately felt on the effect of income and capacity to purchase a diversity of foods.

The way the food is utilized by the body affects the nutritional status of individuals and informs the susceptibility of such individuals to diseases. This is observed as changes in human health that has been affected by new disease patterns created as a result of the response of different pests and vectors to climate change. As climate change increases, there are more waterborne infections and diseases in areas that are prone to flooding. This is because there will be increase in the population of pests and vectors, and new diseases will emerge, which can affect plant's growth and ultimately the food chain. As a result of the diseases, people's physiological capacity to obtain necessary nutrients from the foods consumed will be affected. Therefore, they can become nutritional food insecure despite the supply of food being sufficient (FAO 2008).

Climate Change Impact on Food System

Yields of crops particularly crops grown on an annual basis will fluctuate as a result of the change in rainfall and temperature patterns; these changes will affect the continuous supply of such crops (Al et al. 2008; Ali et al. 2014). Droughts and floods will become more frequent because of climate change, and this will adversely affect stability and availability of food leading to food security. Therefore, in local communities, the food system will be adversely affected by the timing and amount of rainfall in different seasons because most agricultural systems in such communities are rainfed (Al et al. 2008; Ali et al. 2014).

Grain reserves are used to mitigate for crop failures in disaster situations and to sustain food aid programs for displaced persons and refugees. For grain to be kept as reserves, they have to be properly stored or preserved correctly. When humidity and temperature increase because of climate change, some areas become unfit to store the grains, and additional funding will be required to store or preserve them. Invariably because of the higher cost, the ability of many underdeveloped countries to maintain sufficient grain reserves is limited or grain reserves can be inadequate to respond to natural or human disaster (Al et al. 2008; Edame et al. 2011).

Impact of Climate Change Adaptation on Food Security and Nutrition

A significant effect of global climate change is the altering of global rainfall patterns, with certain effects on agriculture (Ali et al. 2014). Climate change will have variable impacts on agriculture based on multiple factors including changes in temperature, precipitation, and humidity (Pitesky et al. 2014). Rainfed

Table 1 The impact of climate change on food security

Climate change	Impacts
Increase in average temperature	Reduced quantity and reliability of agricultural yield
	Increased heat stress in livestock
	Destruction of crops or lowering crop productivity
	Decline in certain fish stocks due to increased sea temperature
Change in amount of rainfall	Reduced water availability for crop and livestock
	Heavy reliance on irrigation
	Poor quality of crops due to deteriorating water quality
Increased severity of drought	Decreased crop yield
	Increased probability of fire
Increased intensity of extreme events	Soil erosion
	Increased land degradation and desertification
	Inability to cultivate land
	Damage to crops and food stores

Source: Masipa (2017)

agriculture constitutes 80% of global agriculture. Many of the 852 million poor people in the world live in parts of Asia and Africa that depend on rainfall to cultivate food crops. Climate change will modify the components of the water balance, namely, rainfall, evaporation, runoff, and soil moisture storage. This means that the full crop water requirement will not be satisfied in such a growing season. Extended drought can cause the failure of small and marginal farms with resultant economic, political, and social disruption, more than is currently observed (Table 1).

Crop Diseases and Effects on Crop Yields

Crop diseases are very important because of their economic impact (Esker et al. 2012), and how they are managed can have a large influence on agricultural productivity (Savary et al. 2012; Carroll et al. 2017). The growing population of the world and the overall reduction in global production of staples such as rice and wheat due to climate change (Ray et al. 2019) are two fundamental challenges that need to be tackled, with crop protection against diseases having an obvious role to play in bringing solutions to these challenges. Crop diseases affect the life cycles of any growing plant. They affect crop production from seedling stage through budding, flowering, and fruiting until harvesting. Crop diseases that occur during seedling and budding stages will in most cases cause the death of the plants as the growing point is damaged. These deaths lead to complete crop failure and high losses. In addition, losses at the time of harvest in most cases cause the reduction in yield in terms of size and quality and thus also a reduction in overall production (Savary et al. 2012). Therefore, loss of crops caused by pests, infections, or even weeds or animal attacks is a risk to farmers, their family or household income, and food security worldwide. Losses of crops due to yield reduction can be defined qualitatively and quantitatively (Cerda et al. 2017) in terms of the reduction in the value of the crop or the financial gain from crop sales.

In order to assess crop loss based on decrease in yield of crop, the variance between the real yield (Y) and the attainable yield that can be obtained (Ya) is taken into consideration as Ya – Y according to the definition of crop loss by Esker et al. (2012). The attainable yield can be achieved based on the different climatic and environmental conditions at specific locations without the impact of pests, pathogens, and weeds as well as other influences that may decrease yields within available production techniques. These climatic conditions among others include temperature, rainfall, solar radiation, and soil composition. The influence of pests, diseases, and other limiting factors in the actual sense and in real-life situation is reflected on the real yield (Esker et al. 2012; Savary et al. 2006).

Plant diseases are not just important because of their impact on crop yield, which is quantitative, but also their impact on the quality of the grain or product. As crops are attacked by disease pathogens, toxins are released into these crop which reduces the quality of the crop and can also cause infection in both animals and humans (Gururani et al. 2012).

Climate Change Impact on the Spread of Plant and Animal Pests and Diseases

Climate change can influence trends in seasons and the prevalence of disease-causing organisms and vectors and therefore also the use of herbicides and fungicides (Boxall et al. 2008). These reactions will vary among crops and geographical location. As a result of climate variability, the use of pesticide is expected to increase in the United States (Chen and McCarl 2001). Nevertheless, the estimated effects of climate change on pesticide use are a function of the type of crop and the geographical location. For example, from the United States, in Kansas and Colorado, the projected use of pesticides on wheat increased by 14% and decreased by 10%, respectively, while that of Illinois differed based on which crops was planted with increases on corn and decreases on soybean by 10% and 3%, respectively (Harrus and Baneth 2005). This shows that when there is an increase in temperature due to climate variability, there could be an outbreak of pests, pathogens, and/or vectors and their hosts (Lake et al. 2012). This outbreak could also lead to the use of more biocides and veterinary medicines in livestock management and eventual resurgence of antibiotic-resistant pathogens in both animal and human populations (Lake et al. 2012; Kemper 2008).

Climatic Conditions Favoring Crop Diseases: Wheat Rust Disease Example

As climate changes, the way disease pathogens adapt to the environment also changes leading to new disease occurrences (Fig. 1). For example, it was observed that warmer weather conditions favor the establishment and progress of wheat stem rust infection (Saunders et al. 2019). In certain regions of the world, during early growing season, high temperatures are experienced. This can cause the wheat plants that mature early to be at risk of the stem rust infection due to not being able to resist the buildup of the fungal inoculum (Saunders et al. 2019). This is because stem rust mainly needs warmer conditions for infection (Saunders et al. 2019). For example, in Germany in 2013, winter wheat production was hindered by a wet winter with low temperatures, and early warm summer temperatures led stem rust infection afterward. (Olivera Firpo et al. 2017).

Climatic Change Impact Postharvest

Over 80% of attainable yield of cotton and over 50% for other crops would be affected by pests and diseases with resulting effects on humankind. These effects which also include spoilage during postharvest and storage are felt more in resource-poor regions (Oerke 2006). It is estimated that an average of 13% of global harvest is lost to pests and diseases, and an average of 9% to postharvest losses especially in developing countries that lack the infrastructures to store harvested crops (Oerke 2006). In some cases, such infrastructures are destroyed due to climate change such as storms, hurricanes, and tornadoes (Agrios 2005). Also, destruction caused as a result of plant diseases can be far reaching and alter the course of society and political history (Chakraborty and Newton 2011).

Storms have led to loss of crops, livestock, and infrastructure and increase in the cost of grains as a result of increase in production cost. This increase also leads to lack of resource power of the minority poor in gaining access to food and invariably increases food insecurity.

The production and quality of essential food crops are hindered by plant diseases, which also affect food quality and safety. Besides reducing yield, they have significant effects on human and animal health. As a result of global climate change, the challenge of mycotoxins and pesticide residues are serious safety issues to be mitigated (Miraglia et al. 2009). Mycotoxins are important as they cause serious health issues because of the interactions between fungus and plant cells, which are also stimulated by conducive weather conditions and the soil type. In some areas like the United States, increases in temperature increased the production of mycotoxins such as aflatoxins, while in other regions like Australia, increase in temperature could be effective in the reduction of fungal growth and invariable mycotoxin production (Russell et al. 2010; Lake et al. 2012).

Climate change may affect the transport of pathogens and chemicals into food from flooding and wind bearing aerosols from contaminated environments and may change the composition of the transported materials (Plumlee et al. 2005; Boxall et al. 2008). This is also observed in the transportation of food crops from the farm before it is consumed where higher temperatures can favor the growth and replication of bacterial population like *Salmonella* spp. leading to food poisoning (Lake et al. 2012).

Climate Change Impact on Microbial Biodiversity

Climate change affects the diversity and types of macro- and microorganisms globally. Biodiversity can be measured at three levels, that is, genetic, species, and ecosystem level. Agriculture and the livelihood of most people globally depend on the genetic diversity of species and the types of species within the ecosystem (ten Kate and Laird 1999). Based on the adverse effect of chemicals in agriculture, and on human health and the environment, the use of chemicals is gradually being phased out to accommodate other techniques and technologies that can control plants and animal's pests and diseases and reduce adverse effects currently experienced by living things and the environment.

In order to achieve agricultural biodiversity, diseases and pests must be dealt with since they are one of the most deadly blows to food security. Seven thousand plant species and several thousand animal species have been used historically for human nutrition and health requirements (Toledo and Burlingame 2006). However, since the 1900s, only 12 crops and 14 animal species now provide most of the world's food, so there is a very real threat to global biodiversity and also food security.

Soil biological management techniques can improve crop yield and quality, especially through controlling pests and diseases and enhancing plant growth. Soil biodiversity determines the resource use efficiency, as well as the sustainability and resilience of low-input agro-ecological systems, which ensure the food security of much of the world's population, especially the poor (Paoletti et al. 2000). Part of the value placed on soil biodiversity is seen in biological control of pests and pathogens that improve plant health and thus also crop production and invariably improving local food security.

Biocontrol of Plant Diseases

Biocontrol is one of the integrated disease management practices that help to reduce yield loss of crops (Thurman et al. 2017). It is the control of plant diseases by applying living biological agents or exudates from such biological agents to the plants. The application could be via foliar application to the leaves or via the roots or seed inoculation prior to planting. Most biocontrol agents (BCA) are either from the bacterial or fungal families with a few from the viral and nematodal families (O'Brien 2017; Ajilogba and Babalola 2019). Some BCAs are either rhizospheric or endospheric microbes, as they attack the pathogens using different modes including phosphate solubilization, antagonism, and hydrogen cyanide production (Ajilogba and Babalola 2016). One important fact about biocontrol is that they are specific to pathogens and sometimes to crops making them very effective. Their effectiveness is also determined by the mode of inoculation, duration of inoculation, time of inoculation, and whether they are applied singly or in a mixture. Climate change also influences biocontrol because changes in the environment affect different dimension of agricultural production. In the context of climate change, with increasing temperature and variable rainfall patterns, changes are also observed in the cycling of nutrients and amount of soil water as a result of decreases in rainfall. The rainfall pattern also affects the patterns of incidence of plant pests, pathogens, and diseases, and invariably food production and security are ultimately affected (Lin 2011; Fuhrer 2003; Jones and Thornton 2003).

Ocean Acidification

This is more of a recent phenomenon with climate change. It is one of the potentially most harmful and deleterious effects of climate change. Over 30% of carbon dioxide produced in the atmosphere is absorbed by sea and oceans leading to the production of acids when carbon dioxide mixes with water. Most marine creatures cannot survive at low pH. Fish production is at a risk both directly and indirectly. The basic producers in the food chain such as planktons are also at risk, and this creates a reduction in food available to the large fish that consume then directly such as tuna, making them at risk of starving. Increasing acidity could harm shellfish by weakening their shells, which are created by removing calcium from seawater (Doney et al. 2014). Acidification also threatens the structures of sensitive ecosystems upon which some fish and shellfish rely (Hatfield et al. 2014; Ziska et al. 2016). Fish is not only a source of proteins and healthy fats but also a unique source of essential nutrients, including long-chain omega-3 fatty acids, iodine, vitamin D, and calcium. Therefore, the effects on climate change on the acidity of the oceans will cause a decrease in annual fish caught and processed at a good source of protein for humankind. Particularly in the Asian-Pacific region, fish protein forms a vital component of the diet (Needham and Funge-Smith 2014). It is not only true for small island states, with their strong dependence upon fisheries; it was also found in continental Asian countries even far inland due to large rivers, waterways, and many rice paddies. Therefore, the importance of fish in the diet should not be overlooked in discussions on nutrition and food security and expected changes under the climate change conditions.

Conclusions

Climate change is a reality, both at global and the regional scales. The magnitude of change is still uncertain and highly variable, as is the timing of the change. However, the severe nature of the impacts which climate change could have on all sectors of Africa's economy is becoming clear. The ability of individuals and communities to adapt to climate change depends on their vulnerability, exposure, and adaptive capacity. In turn, this is related to their financial and social capital, such as social networks and community cohesion.

Recommendations

Adaptation is a key factor that will shape the future severity of climate change impacts on food production; therefore, consideration of additional food such as underutilized crops and/or neglected foods can play a role in increasing food and nutrition supply.

There is, therefore, a need for an integrated policy approach to protect the arable land.

Further assistance needs to be given to small-scale farmers in rural areas in order to strengthen their crop production.

The restoration of forestation should be encouraged in order to make adequate use of forests to address climate change.

Globally and whenever necessary, nations should specifically relate their mitigation and adaptation initiatives to their policy decisions on environmental sustainability in agriculture industry. These policies should both be holistic and anti-poor in order to bridge the gap on the unequal effect that climate change may have on the different agricultural industries.

Consideration should be given to the economic, institutional, and social impacts of climate change by making policies that will target the global poor and vulnerable especially gender based in order to help them create communities that are resilient and robust.

Disasters caused by climate change in agriculture communities demand immediate emergency relief response and comprehensive facilities such as shelters especially in the animal farming communities. Such developments frequently involve initiatives that will provide opportunities for people to develop societies and cooperatives that are stronger and more productive and thereby becoming more self-reliant.

Adaptation and mitigation measures and strategies for climate change need to be designed and implemented through widely understood processes. Affected agricultural communities, organizations, the poor, and other vulnerable groups must be allowed to participate as stakeholders.

Global and national relevant indicators to ensure the participation of the poor and vulnerable should be developed; this could be done by keeping gender-stratified data to predict and determine the impacts of climate change on them and also their geographic locations.

Other recommendations include availability of suitable technology and decision support tools, removal of institutional barriers, and need for basic research to increase knowledge and information.

Conflict of Interest

Authors have no conflicts of interest to report.

Acknowledgments The authors want to thank the National Research Foundation for the grant to the first author and the Agricultural Research Council – Soil, Climate, Water (ARC-SCW) for logistics.

References

Agrios G (2005) Plant pathology, 5th edn. Elsevier Academic Press, London

Ajilogba CF, Babalola OO (2016) RAPD profiling of *Bacillus* spp. with PGPR potential and their effects on mineral composition of tomatoes. J Hum Ecol 56(1–2):42–54

Ajilogba CF, Babalola OO (2019) GC–MS analysis of volatile organic compounds from Bambara groundnut rhizobacteria and their antibacterial properties. World J Microbiol Biotechnol 35(6):83

Al W, Orki G, Clima O (2008) Climate change and food security: a framework document. FAO, Rome

Ali F, Thaver I, Khan SA (2014) Assessment of dietary diversity and nutritional status of pregnant women in Islamabad, Pakistan. J Ayub Med Coll Abbottabad 26(4):506–509

Allen LH Jr (1991) Effects of increasing carbon dioxide levels and climate change on plant growth, evapotranspiration, and water resources. In: Managing water resources in the west under conditions of climate uncertainty. Proceedings of a colloquium, 14–16 Nov 1990, Scottsdale, AZ. National Academy Press, Washington, DC, pp 101–147

Bierbaum R, Stults M (2013) Adaptation to climate change. Mich J Sustain 1. https://doi.org/10.3998/mjs.12333712.0001.004

Bierbaum R, Smith J, Lee A, Blair M, Carter L, Stuart Chapin F III, Fleming P, Ruffo S, McNeeley S, Stults M, Wasley E, Verduzco L (2013) A comprehensive review of adaptation in the United States: more than before, but less than needed. Mitig Adapt Strateg Glob Chang 18(13):361–406. https://doi.org/10.1007/s11027-012-9423-1

Bonan GB (2008) Forests and climate change: forcings, feedbacks, and the climate benefits of forests. Science 320(5882):1444–1449

Boone RB, Conant RT, Sircely J, Thornton PK, Herrero M (2018) Climate change impacts on selected global rangeland ecosystem services. Glob change Biol 24(3):1382–1393

Boxall AB, Hardy A, Beulke S, Boucard T, Burgin L, Falloon PD, Haygarth PM, Hutchinson T, Kovats RS, Leonardi G (2008) Impacts of climate change on indirect human exposure to pathogens and chemicals from agriculture. Environ Health Perspect 117(4):508–514

Brida A-B, Owiyo T, Sokona Y (2013) Loss and damage from the double blow of flood and drought in Mozambique. Int J Glob Warming 5(4):514–531

Carroll CL, Carter CA, Goodhue RE, Lawell C-Y (2017) Crop disease and agricultural productivity. National Bureau of Economic Research, Cambridge, MA

Cerda R, Avelino J, Gary C, Tixier P, Lechevallier E, Allinne C (2017) Primary and secondary yield losses caused by pests and diseases: assessment and modeling in coffee. PLoS One 12(1): e0169133

Chakraborty S, Newton AC (2011) Climate change, plant diseases and food security: an overview. Plant Pathol 60(1):2–14

Chen X (2015) Climate change: causes, impacts and adaptation. Project supported by Govenor State University Intellectual Life Grant, pp 1–9

Chen C-C, McCarl BA (2001) Pesticide usage as influenced by climate: a statistical investigation. Clim Chang 50(1–2):475–487

Doney S, Rosenberg AA, Alexander M, Chavez F, Harvell CD, Hofmann G, Orbach M, Ruckelshaus M (2014) Oceans and marine resources. In: Climate change impacts in the United States. The third national climate assessment. U.S. Global Change Research Program, Washington, DC. https://doi.org/10.7930/J0RF5RZW

Du Toit A, Ziervogel G (2004) Vulnerability and food insecurity: background concepts for informing the development of a national FIVIMS for South Africa. Background paper for FIVIMS-ZA consortium. http://www.agis.agric.za/agisweb/agis.html

Edame EG, Ekpenyong AB, Fonta WM, Duru EJC (2011) Climate change, food security and agricultural productivity in Africa: issues and policy directions. Int J Humanit Soc Sci 1(21):205

Esker PD, Savary S, McRoberts N (2012) Crop loss analysis and global food supply: focusing now on required harvests. CAB Rev 7(052):1–14

FAO (2003) Focus on Food Insecurity and Vulnerability – A review of the UN System Common Country Assessments and World Bank Poverty Reduction Strategy Papers. FIVIMS Secretariat and Wageningen University and Research Centre. https://www.fao.org/DOCREP/006/Y5095E/Y5095E00.htm

FAO (2008) Climate change and food security: a framework document. Food and Agricultural Organization of United Nation, Rome, Italy

Food and Agriculture Organization (FAO) (1996) Rome Declaration on world food security. FAO, Rome

Food and Agriculture Organization (FAO) (2015) The impact of natural hazards and disasters on agriculture and food security and nutrition: a call for action to build resilient livelihoods. Food and Agriculture Organization of the United Nations, Rome

Food and Agriculture Organization (FAO) (2016) Climate change and food security: risks and responses. Food and Agriculture Organization of the United Nations, Rome

Fuhrer J (2003) Agroecosystem responses to combinations of elevated CO_2, ozone, and global climate change. Agric Ecosyst Environ 97(1–3):1–20

Gerber PJ, Steinfeld H, Henderson B, Mottet A, Opio C, Dijkman J, Falcucci A, Tempio G (2013) Tackling climate change through livestock: a global assessment of emissions and mitigation opportunities. FAO, Rome

Gururani MA, Venkatesh J, Upadhyaya CP, Nookaraju A, Pandey SK, Park SW (2012) Plant disease resistance genes: current status and future directions. Physiol Mol Plant Pathol 78:51–65

Harrus S, Baneth G (2005) Drivers for the emergence and re-emergence of vector-borne protozoal and bacterial diseases. Int J Parasitol 35(11–12):1309–1318

Hatfield J, Takle G, Grotjahn R, Holden P, Izaurralde RC, Mader T, Marshall E, Liverman D (2014) Agriculture. In: Climate change impacts in the United States. The third national climate assessment. U.S. Global Change Research Program, Washington, DC

ILO (2007) Employment by sector. In: Key indicators of the labour market (KILM), 5th edn. ILO, Geneva

IPCC (2007) The physical science basis. Contribution of working group I to the fourth assessment report of the Intergovernmental Panel on Climate Change. Cambridge University Press, Cambridge, UK. 996pp

IPCC (2012) Managing the risks of extreme events and disasters to advance climate change adaptation. In: CB Field, V Barros, TF Stocker, et al., editors. A Special Report of Working Groups I and II of the Intergovernmental Panel on Climate Change. Cambridge University Press, Cambridge, NY, USA

IPCC (2014) Climate change 2014, impacts, adaptation, and vulnerability, Part A: global and sectoral aspects, working group II contribution to the fifth assessment report of the Intergovernmental Panel on Climate Change. Cambridge University Press, Cambridge, UK

Jones PG, Thornton PK (2003) The potential impacts of climate change on maize production in Africa and Latin America in 2055. Glob Environ Chang 13(1):51–59

Kambrekar D, Guledagudda S, Mohankumar A (2015) Impact of climate change on insect pests and their natural enemies. Karnataka J Agric Sci 28:814–816

ten Kate K, Laird SA (1999) The commercial use of biodiversity: access to genetic resources and benefit-sharing. Earthscan, London

Kemper N (2008) Veterinary antibiotics in the aquatic and terrestrial environment. Ecol Indic 8 (1):1–13

Lake IR, Hooper L, Abdelhamid A, Bentham G, Boxall ABA, Draper A, Fairweather-Tait S, Hulme M, Hunter PR, Nichols G, Waldron KW (2012) Climate change and food security: health impacts in developed countries. Environ Health Perspect 120(11):1520–1526. https://doi.org/10.1289/ehp.1104424

Leff B, Ramankutty N, Foley JA (2004) Geographic distribution of major crops across the world. Glob Biogeochem Cycles 18(1):GB1009. https://doi.org/10.1029/2003GB002108

Lin BB (2011) Resilience in agriculture through crop diversification: adaptive management for environmental change. Bioscience 61(3):183–193

Little A (2019) The fate of food what we'll eat in a bigger, hotter, smarter world. Libr J 144(5):138

Masipa TS (2017) The impact of climate change on food security in South Africa: current realities and challenges ahead. Jàmbá 9(1):1–7

Miraglia M, Marvin H, Kleter G, Battilani P, Brera C, Coni E, Cubadda F, Croci L, De Santis B, Dekkers S (2009) Climate change and food safety: an emerging issue with special focus on Europe. Food Chem Toxicol 47(5):1009–1021

Moore B, Allard G (2008) Climate change impacts on forest health: insect pests, diseases and invasive alien species. In: International conference on adaptation of forests and forest

management to changing climate with emphasis on forest health: a review of science, policies and practices, p 170
Needham S, Funge-Smith SJ (2014) The consumption of fish and fish products in the Asia-Pacific region based on household surveys. FAO Regional Office for Asia and the Pacific, Bangkok, Thailand. RAP Publication 2015/12. 87pp
O'Brien PA (2017) Biological control of plant diseases. Australas Plant Pathol 46(4):293–304
Oerke E-C (2006) Crop losses to pests. J Agric Sci 144(1):31–43
Olivera Firpo P, Newcomb M, Flath K, Sommerfeldt-Impe N, Szabo L, Carter M, Luster D, Jin Y (2017) Characterization of *Puccinia graminis f.* sp. *tritici* isolates derived from an unusual wheat stem rust outbreak in Germany in 2013. Plant Pathol 66(8):1258–1266
Paoletti MG, Dufour DL, Cerda H, Torres F, Pizzoferrato L, Pimentel D (2000) The importance of leaf-and litter-feeding invertebrates as sources of animal protein for the Amazonian Amerindians. Proceedings of the Royal Society of London. Series B: Biol Sci 267(1459):2247–2252
Pitesky M, Gunasekara A, Cook C, Mitloehner F (2014) Adaptation of agricultural and food systems to a changing climate and increasing urbanization. Curr Sustain Renew Energy Rep 1:43. https://doi.org/10.1007/s40518-014-0006-5
Plumlee GS, Foreman WT, Griffin DW, Lovelace JK, Meeker GP, Demas CR (2005) Characterization of flood sediments from hurricanes Katrina and Rita and potential implications for human health and the environment. In: Science and the storms: the USGS response to the hurricanes of 2005. US Geological Survey, Denver, pp 245–256
Porter JR, Semenov MA (2005) Crop responses to climatic variation. Philos Trans R Soc B Biol Sci 360(1463):2021–2035
Pragna P, Chauhan SS, Sejian V, Leury BJ, Dunshea FR (2018) Climate change and goat production: Enteric methane emission and its mitigation. Animals 8(12):235. https://doi.org/10.3390/ani8120235
Pretty JN, Ball AS, Lang T, Morison JI (2005) Farm costs and food miles: an assessment of the full cost of the UK weekly food basket. Food Policy 30(1):1–19
Ray D, West PC, Clark M, Gerber JS, Prishchepov AV, Chatterjee S (2019) Climate change has likely already affected global food production. PLoS One 14(5):e0217148. https://doi.org/10.1371/journal.pone.0217148
Rojas-Downing MM, Nejadhashemi AP, Harrigan T, Woznicki SA (2017) Climate change and livestock: impacts, adaptation, and mitigation. Clim Risk Manag 16:145–163
Russell R, Patterson M, Lima N (2010) How will climate change affect mycotoxins in food? Food Res Int 43:1902–1914
Rust JM, Rust T (2013) Climate change and livestock production: A review with emphasis on Africa. S Afr J Anim Sci 43:255–267
Sage C (2014) Impacts of climate change on food accessibility. In Handbook of Global Environmental Change Springer Science + Business Media, pp 709–715
Saunders DG, Pretorius ZA, Hovmøller MS (2019) Tackling the re-emergence of wheat stem rust in Western Europe. Commun Biol 2(1):51
Savary S, Teng PS, Willocquet L, Nutter FW Jr (2006) Quantification and modeling of crop losses: a review of purposes. Annu Rev Phytopathol 44:89–112
Savary S, Ficke A, Aubertot JN, Hollier C (2012) Crop losses due to diseases and their implications for global food production losses and food security. Food Secur 4(4):1–19
Simpson BM (2017) Preparing smallholder farm families to adapt to climate change. Managing crop resources. Pocket guide 2. Catholic Relief Services, Baltimore
Thornton PK (2010) Livestock production: recent trends, future prospects. Philos Trans R Soc Lond B Biol Sci 365(1554):2853–2867. https://doi.org/10.1098/rstb.2010.0134
Thurman JH, Crowder DW, Northfield, TD (2017) Biological control agents in the Anthropocene: current risks and future options. Curr Opin Insect Sci 23:59–64. https://doi.org/10.1016/j.cois.2017.07.006
Toledo A, Burlingame B (2006) Biodiversity and nutrition; a common path toward global food security and sustainable development. J Food Compost Anal 19:477–483

Tubiello FN, Amthor JS, Boote KJ, Donatelli M, Easterling W, Fischer G, Gifford RM, Howden M, Reilly J, Rosenzweig C (2007) Crop response to elevated CO_2 and world food supply: a comment on "Food for thought..." by Long et al., Science 312: 1918–1921, 2006. Eur J Agron 26(3):215–223

Van der Meer PJ, Jorritsma ITM, Kramer K (2002) Assessing climate change effects on long-term forest development: adjusting growth, phenology, and seed production in a gap model. Forest Ecol Manag 162(1):39–52

Yadav B, Gynendra S, Verma AK, Dutta N, Sejian V (2013) Impact of heat stress on rumen functions. Vet World 6:992–996

Zewdie Y (2003) Forest access: policy and reality in Kafa, Ethiopia. Leisa-leusden 19:21–22

Ziervogel G Ericksen, PJ (2010) Adapting to climate change to sustain food security. Wiley Interdiscipl Rev: Climate Change 1(4):525–540

Ziervogel G, Frayne B (2011) Climate change and food security in Southern African cities. Urban food security series. Queen's University and AFSUN, Kingston/Cape Town

Ziska L, Crimmins A, Auclair A, DeGrasse S, Garofalo JF, Khan AS, Loladze I, Pérez de León AA, Showler A, Thurston J, Walls I (2016) Food safety, nutrition, and distribution. In: The impacts of climate change on human health in the United States: a scientific assessment. U.S. Global Change Research Program, Washington, DC

2

Smart Climate Resilient and Efficient Integrated Waste to Clean Energy System in a Developing Country: Industry 4.0

Anthony Njuguna Matheri, Belaid Mohamed and Jane Catherine Ngila

Contents

A. N. Matheri (✉) · B. Mohamed
Department of Chemical Engineering, University of Johannesburg, Johannesburg, South Africa
e-mail: mbelaid@uj.ac.za

J. C. Ngila
Department of Chemical Science, University of Johannesburg, Johannesburg, South Africa

Academic Affair, Riara University, Nairobi, Kenya
e-mail: jcngila@riarauniversity.ac.ke

Abstract

Climate change impacts a natural and human system on the entire globe. Climate-related extreme weather such as drought, floods, and heat waves alters the ecosystems that society depends on. Climate, land, energy, and water systems (CLEWS) are a critical aspect of high importance on resource availability, distribution, and interconnection. The nexus provides a set of guidelines to South Africa that aims on creating a level playing field for all sectors while achieving the aims of the SDGs that are cross-sectoral and multilevel approaches to climate change. The nexus expressed three domains that included resources, governance, and security. It integrated a smart climate resilient with inclusion of the governance and involvement of the stakeholders. Recognition of spatial and sector interdependencies should inform policies, investment and institutional for enhancing nexus security and climate change towards making transition green carbon deals. The nexus offers an integrated approach that analyzes the trade-offs and synergies between the different sectors in order to maximize the efficiency of using the resources that adapt institutional and optimum policy arrangements. Economic transformation and creation of employment through green economy is one of the COP26 green deal agendas in curbing the carbon emissions (green house emission, industrial processes, fuel combustion, and fugitive emissions) as mitigation to climate change, which is cost-effective and economically efficient. The future climate change policy in the developing countries is likely to be both promoted by climate technology transfer and public-private cooperation (cross-sector partnership) through the technology mechanism of the nexus and inclusion of the gender.

Keywords

Adaptation · CLEWS · Climate change · Climate technology · Ecosystem · Green economy · Policy · Resilience · SDGs · Nexus

Introduction

Green revolution, disruption of the natural resources and social-economic impact are becoming increasingly in Industry 4.0 (Fourth Industrial Revolution-41R). The is due to enhancement of the adaptive capacity for the complex global challenges in advanced nexus approach to the sustainable management of climate change, land use, water, energy, and food. This serve the paradigmatically that was isolated and understanding of the interrelation of the WEF, human well-being, resilient ecosystem, climate change that coexists within the planetary boundaries. The shift in society disruption legion of unfortunate solution to an environment or development challenges that end up creating new, unforeseen problems and dilemma (Newell et al. 2019; Matheri et al. 2020). The nexus describes the interconnections and interdependencies between the land, water, energy, and food (LWEF) sectors. These interdependencies of the LWEF securities have received growing attention in the past years by researchers and policy-makers. Lack of understanding of the nexus has been described as a major global economic challenge by the World Economic Forum. This put forward the nexus approach as a fundamental shift for sustainable development (Hassan et al. 2018). A review on the LWEF and land utilization needs to be addressed on policy-making and decision-making concerning the nexus framework and climate change. This bridges the gap between science and policy in the implementation of nexus. More importantly, the nexus can also support achieving the United Nations Sustainable Development Goals (SDGs), 2030 Agenda, because of its close relation to three SDGs: SDG2 "zero hunger," SDG6 "clean water and sanitation for all," SDG7 "affordable and clean energy," and SDG13 "climate action" (Tashtoush et al. 2012).

The nexus is expressed into three domains that include resources, governance, and security. The domain requires inclusion of the policy and institution expertise, natural resources use, environmental science, and engineering expertise for the mutual perspective of multiple decision-making, solution, and resource recovery (Scott et al. 2015, 2018). The priorities areas for climate change include decarbonization of the economy, digitization, decentralization of the production, transition of 100% renewable energy, carbon taxation (tax the polluters not people), access to sustainable finance, assist small island and least developing countries, support people affected by climate change, increase accessibility and availability of the jobs and livelihood, nature-based solutions, stronger commitments by the major emitters, and commitments to achieve carbon neutrality by 2050 (United Nation (UN) 2020). The focus is placed on implementation of the nexus concepts with a smart climate resilient on nexus regional dialogue programs by United Nations (UN) bodies, World Bank, GIZ, European Union (EU), Africa Union (AU), Middle East and North Africa (MENA), and World Economic Forum among others.

Paris Agreement on Climate Change

Climate change results in disruption of national economies that affect the countries, communities, living standards, and flora and fauna. With likely shift of the climate zones, long-term, changes of the rainfall patterns, and raising of the temperature, climate change shock is expected to increase frequency putting pressure in energy, food, water supply, and competition of land ownership in unaffected areas (Stocker et al. 2013; Carter and Gulati 2014). The historic Paris agreement provides countries with strong global climate change response by keeping the temperature rise below 2–1.5 °C. It encourages parties and stakeholders to reduce the impacts of the climate change (ambition and implementation to ensure highest mitigation and adaptation efforts by all parties with building climate resilience).

The intergovernmental climate change panels have recorded the following:

- 1901–2010: Warm oceans and diminishing snows and ice by 1.07 million km^2 with rise of the sea levels by 19 cm
- 1880–2012: Increase of the temperature by 0.85 °C due to warmer climate
- 1990 to date: Global emission of the CO_2 that has increased by 50%

The agreement solidifies international cooperation on investment of a low-emission economy (green economy) that contains transparency framework, mutual trust, and confidence. Industry 4.0 is overseeing the scalable, affordable solutions that are more resilient to the economy and mostly helping the developing countries toward the low-carbon economy. Major technologies and institutions will see a shift of containing and ensuring that global warming does not exceed the threshold. SDG 13 defines the commitment by strengthening resilience and adaptive capacity; integrating climate change strategies, policies, and planning; creating awareness on adaptation, mitigation, and impact reduction; and implementing meaningful mitigation again and transparency through UNFCCC (United Nation Framework Convention on Climate Change) framework and inclusion of the marginalized communities and planning and management (SDGs, UN 2019). The impact of the climate change will be amplified through interdependence and interconnection among the resources, land, water, energy, and food.

Climate-Land-Water-Energy-Food Nexus

The global web of mutual interlinkages defines the climate-land-water-energy-food nexus on societal changes that drive growth and demand. The ongoing disruption of the environment is likely to alter the accessibility or availability of land, energy, water, and food that is central to climate change policies and natural resource management (Markantonis et al. 2019). Holistic and integrated approaches to resource planning and management have been largely embraced by decision-makers and stakeholders although the benefits of the nexus approach may appear obvious to its advocates. The nexus is related to integrated management, consumption, economic resources, policy, security, and approach. The nexus concept needs to be approached beyond the research and development domain.

The WEF nexus has emerged as a useful disruption in Industry 4.0 in understanding multiple interdependencies that coexist between land use, water, energy, food, and climate change. It is a multidisciplinary that cuts across both state and non-state sectors. The WEF has potential to unlock groundbreaking solutions to complex problems with appropriate models and climate change mitigation. The is expressed as National Development Plans (NDP) and United Nations Sustainable Development Goals (SDG) that are emerging in the international agenda at the World Economic Forum on understanding the link between use of resources in providing universal basic rights and climate protection (Seeliger et al. 2018). This was welcomed by the Conference of Parties (COP) by the UNFCCC on

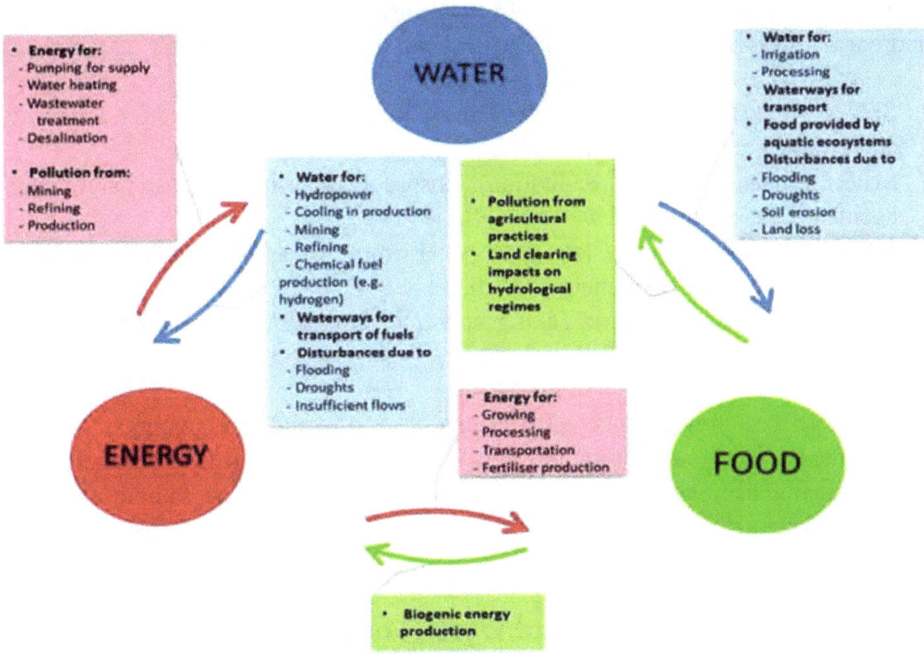

Fig. 1 The water-energy-food nexus concept. Source: IW: LEARN, Germany 2014

enhancement of climate technologies' development and knowledge transfer through the technology mechanisms of greater public-private cooperation (cross-sector partnerships). This is through the action on climate and SDGs, climate technology, education, youth, gender, having open-source information and intellectual properties, climate finance, adaptation and resilience, and capacity building (Forsyth 2007; Forsyth 2005; COP25 2019a, b). Figure 1 shows the WEF nexus concept.

The nexus can provide technology, innovation, and analytical framework by complementing design principles and new policy perspectives for the CLEWS' security policy. Building future planned actions on nexus needs advanced research and development (R&D) such as cost action and Horizon 2020. The role of the WEF nexus is gaining increasing attention in the region from developing countries, which is a partner in the Nexus Regional Dialogues Programme (NRDP) with an aim to create an enabling environment that drives implementation and cross-sectoral engagement of nexus investment projects. The intention is to provide decision-makers with prospects where the theory of the nexus concept has been made operational.

The C40 cities program seeks to build and identify opportunities and build evidence benefits of climate action synergies within the climate, water, and energy nexus. C40 cities have built a platform of networks, adaptation, communications, finance, and other programs on taking bold climate actions that are healthier and more sustainable to resilient cities (C40 2020).

Specific to Agenda 2030, 4 of the 17 SDGs are directly related to the food, water, and energy sectors (Tashtoush et al. 2012; United Nation (UN) 2015):

- SDG 2 (Zero Hunger): End hunger, achieve improved nutrition and food security, and promote sustainable agriculture.
- SDG 6 (Clean Water and Sanitation): Ensure availability and sustainable management of sanitation and water for all.
- SDG 7 (Affordable and Clean Energy): Ensure access to reliable, affordable, sustainable, and modern energy for all.
- SDG 13 (Climate Action): Taking action to combat climate change and the impacts.

Although these four goals directly relate to the individual areas of climate change, water, energy, and food security, progress in 12 of the 17 SDGs is directly related to the sustainable use of resources. Some goals cannot be achieved without a holistic view of the nexus.

Energy Security and Pursuit of Water, Food, and Earth System Resilience

Energy security exists where there is uninterrupted availability of energy sources and distribution at an affordable price to the consumers. With frequent power rationing (load shedding) in many African countries that ranges in hours, this remains a dream in achieving adequate energy production, stable tariffs, relax policies on independent power production (IPP), political stability, intelligent distribution of energy, green buildings, smart metering, energy auction, bidding and pricing, self-driving cars, electronic cars, energy storage, smart cars, high-speed trains, next-generation GPS devices, autonomous vehicles, gyroscopic vehicles, smart roads, hyperloops, micro-mobility, intelligent electric vehicle networks, etc. All of this lies on policies and politics.

The food challenges from energy perspectives include increasing food cooling systems; energy-intensive farm operations; local food chains that is minimizing transport energy; extended crop seasons; artificial intelligence and blockchain technology on food tracking systems; use of remote sensing technology (i.e., drone use for mapping, irrigation, and spraying) to monitor food production, spraying and irrigation; use of hydroponic and aeroponics for urban farming; use of drone to monitor crop production and pest controls; use of robotics technology in farming; and use of artificial intelligence in detecting crop diseases at an early state which all needs intensive use of energy and technology literacy.

The water challenges from energy perspectives include energy intensive of desalination, water reuse, rising demand for carbon-free energy systems, climate change raises water needs of energy, water allocation to energy generation, water capture from atmosphere (humid ambient air) using solar-powered devices, and

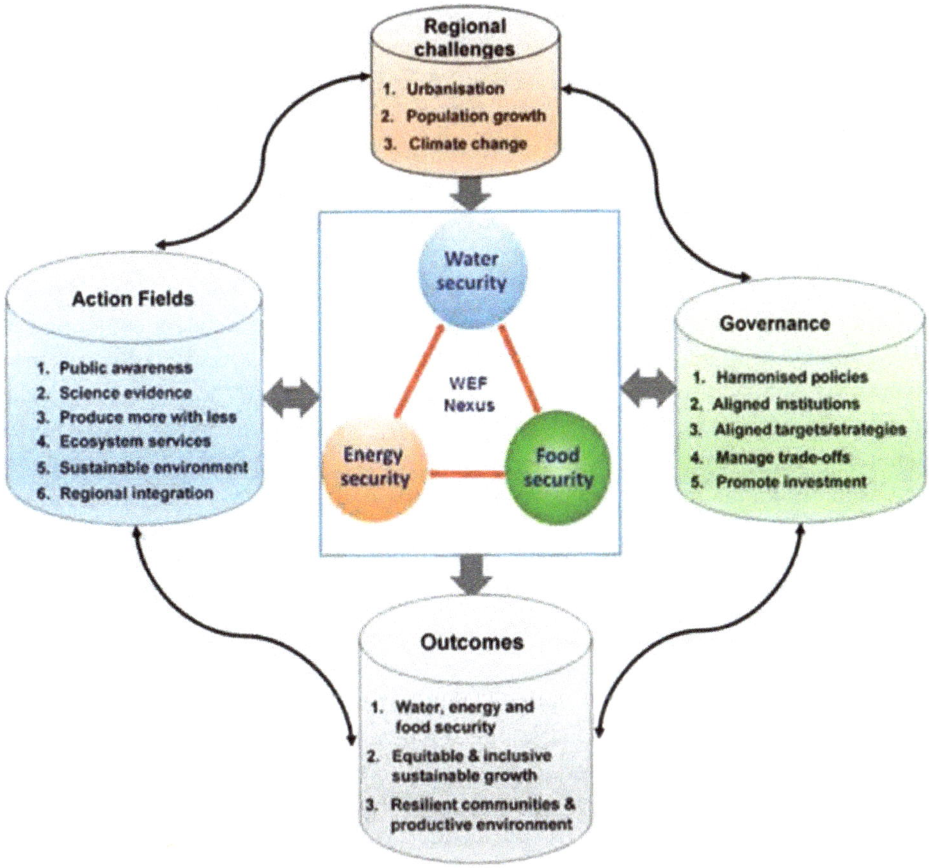

Fig. 2 Conceptual framework of the regional water-energy-food nexus action plan

decentralized wastewater treatment (online wastewater treatment and monitoring using sensors).

The future of transportation lies on our behavior, culture, taboos, and way of life. How often one drives the vehicles, the mode of transport we take, the types of fuels we use, an investment in public transportation, compensation due to shift and investment on green energy (green funding), congestion charges, and more vehicle-pooling and ride-hailing services justify dependence toward zero emission and thus mitigation to climate change. Figure 2 shows the conceptual framework of the water-energy-food action plan (Nhamo et al. 2018; Chirisa and Bandauko 2015).

Reliable Integrated Waste Management and Nexus

The populace development, urbanization, economic advancement, and improvement in expectations for everyday comforts in disruption of Industry 4.0 (4IR) have increased the amount of the waste generation in the cities and reintroduction of the emerging contaminants into waste streams. These wastes pose sanitary, health hazard, and environmental risks. These contaminants end up in water bodies and landfills, prompting to pollution of the entire environment, thus putting a high strain on social, economic, health sector, and climate change. Tackling integrated waste management disruption and energy security requires stipulated regulation and policies that coordinate waste administration frameworks. The future of digital disruption and transformation of industrial platform, energy-water-food generation lies on the decarbonation, decentralization, electrification, and digitalization (sustainable industrialization and diversification in the digital era). Technologies in the era of digital transformation are becoming commercially viable to the integrated systems and services that enable sustainable management and efficient integrated waste management (IWM), energy production, and use. Reliable IWM and data provide an all-inclusive resource for a critical, comprehensive, and informative evaluation of IWM options in all integrated waste management programs. Optimized programmed operations using predictive analytics (i.e., convectional and AI modelling), big data, blockchain, e-citizens, fintech, and data mining are the fundamental apparatuses generally used to assess the policy impact and technology of savvy arrangements, just as to design the most ideal approaches to move from current to more intelligent urban areas. There are several obstacles confronting municipal solid waste management within the cities. Some of such obstacles are interrelation of urbanization and economic growth; change of living standards that causes complexity of the waste stream; overstretching of the superannuated infrastructure; lack of location and facilities to expedite waste separation at source; Intelligent Network Infrastructure (tracking collectors, IoT bins, automated recycler with incentive to users, available recycle plants); and integrated waste management technologies that are handy and costly compared to landfilling and composting. Detachment of waste at the source and embracing zero waste financial motivation urge a family to diminish squander.

The organic waste can be changed over to vitality utilizing waste to vitality elective courses that incorporate transformation, gasification, combustion, pyrolysis and liquefaction, organic procedures, aging, hydrolysis, and anaerobic assimilation for biogas and biomethane creation. An investigation waste quantification to assess the sustainability, characterization to assess the composition, and anaerobic digestion to assess the amount and quality of energy (CH_4) generated by the City of Johannesburg were carried by a team of researchers (University of Johannesburg). The outcomes indicated that 1,444,772 tons per annum of local waste were created in the city of six million residents consistently as announced by the City of Johannesburg (CoJ), South Africa Pikitup (2017). Littering alone costs the city 5.7 million dollars every year, while illicit dumping costs another 6.2 million dollars for each annum. Organic waste can be changed over to vitality utilizing waste or vitality elective courses that incorporate transformation, gasification, combustion, pyrolysis and liquefaction, organic procedures, aging, hydrolysis, fermentation, and anaerobic assimilation for biogas and biomethane creation. Recyclable squander (paper/paperboards, plastics, and glass) was the second biggest part 12%, 19%, and 9%, respectively (Matheri et al. 2018a, b; Matheri 2016, 2020; Fig. 3).

The generated biomasses had potential of the alternative clean fuel production to meet the energy security and climate change mitigation. The physiochemical properties of the biomass showed the energy value equivalent to natural gas. The generated energy will contribute to a reliable, affordable, carbon-neutral, sustainable

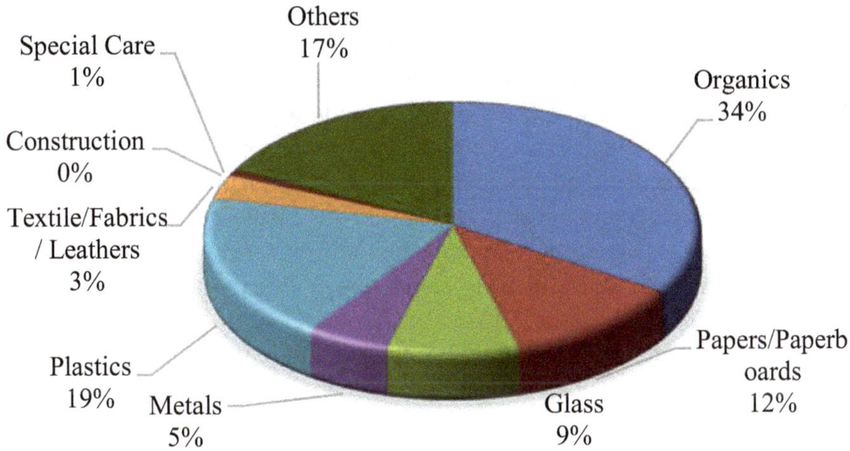

Fig. 3 Quantification of waste generated by the City of Johannesburg, South Africa

form of modern energy. This will help in the development of the conservation of biodiversity, waste management, poverty eradication, and adequate waste to energy recovery management strategies and improve living standards of the citizens (Matheri 2016). The energy system disruption will assist to bridge the gap in the implementation of the SDGs and fourth industrial revolution. The high number of plastic calls for the introduction of the bioplastics to protect the flora and fauna. This adds up pressure on the added value recycling and reduction of waste to landfills. The climate change future policy in developing countries is likely to be both promoted by climate technology transfer and public-private cooperation (cross-sector partnership) through the technology mechanism of the nexus and inclusion of the gender.

Water Security and Pursuit of Energy, Food, and Earth System Resilience

Water is inseparably linked to food and energy production. The interdependence of energy on water is seen in transport, extraction, power generation, irrigation of biofuel crops, processing of fossil fuels, fuel cells, energy storage, etc. (Birol 2012). The emerging issues are cultivated on the regions that are facing adoption of the appropriate water management policies, water scarcity, and approaches fostering the allocation and sustainable use of resources while promoting economic growth. Due to human activities and availability of the water resources, Africa is one of the most vulnerable regions. In order to maximize the efficiency of using the resources, nexus is offering an integrated approach that analyzes the synergies and trade-offs between different sectors whereas adapting optimum institutional and policy arrangements (Markantonis et al. 2019). Lifecycle approaches are widely used by the industry for the assessment of the risk water management. The absence of lifecycle database to complement developing countries' databases is a challenging issue that extensively improved the national accounting system. More positive economic opportunities are created in desalination technologies, smart metering, leak detection (prevent water loss), online water treatment, sensor technology in parameters detection, and smart use of ecosystems to collect and store water and carbon. Water security encourages the population to increase capacity of sustainable water access with adequate quality and quantities for human well-being, sustaining livelihoods and socioeconomic development. This ensures the protection against water-borne diseases, emerging contaminants, and water-related disasters by preserving a climate with peaceful ecosystems and political stability (Hassan et al. 2018; Matheri 2020; UN Water 2013).

The food challenges from water perspectives include raises irrigation demands, diminishing institutional that influences the irrigation schemes, groundwater pumped variables, production shifts poleward, higher elevation, water and wastewater treatment and reuse of the resources (Scott et al. 2018). The energy challenges from water perspectives include low water footprint solar PV and wind, dry cooled thermogeneration, water footprint of multiple energy portfolios, energy generation degrades water quality, energy generation from water resources (hydropower, fuel cells, anaerobic digestion), and energy storage (Scott et al. 2018). Construction of the decentralized solar-powered desalination systems in the dry areas will be one of the game changer deployments of the sustainable technology solution in disadvantaged communities in developing countries. This helps bring millions of liters and address perennial water shortage and thus reduction of water-borne diseases.

Water Pricing

Water pricing currently is often underestimated which is fundamentally of importance in the economic issues that affect the implementation of the nexus on water use. Water pricing is an economic instrument which efficiency depends on the designed and implementation of the WEF nexus framework and investment. This is with regard to the policy choice, economic pursuit, realities, environmental, social, opportunity cost, and cultural cost. Developing countries' adaptation of the EU members' states law and policies on water pricing is one of the key goals in sustainable development (Markantonis et al. 2019).

Food Security and Pursuit of Water, Energy, and Earth System Resilience

Food production needs energy, productive land, and water to grow crops, process food, and maintain livestock. Organic fraction of waste can also be put into value by converting and generating energy via anaerobic digestion and fermentation processes, while other technologies go more to gasification and direct combustion to generate energy and organic fertilizers (Matheri 2020). Such bidirectional links are complicated by the specific external factors that modify the chemical and physical characteristics and composition of water flows. Food consumption (changing dietary habits) and generation of the food waste have large effects on the nexus. This is yet to be accounted in SDGs and cannot be treated with isolation from one another and alternative in the attainable chosen pathways in the water and food systems. Food security exists when everyone, at all times, has economic and physical access to adequate nutritious, safe, and sufficient food that meets their specific dietary preference and needs for an healthy and active life (Hassan et al. 2018). The water challenges from food perspectives include land degradation (salinization), water use, wastewater use for food production, supplemental irrigation of rainfed land, high water footprint of agriculture, and ensuring water allocation to irrigation. The energy challenges from food perspectives include mitigate hydropower-farming trade-offs, energy intensification of agriculture, energy intensification of the biofuel production and food transport, and competition between energy production and food security. One of the futures of food in climate change is podcast on the plant-based meat change to counteract the intensive use of energy.

Biophysical and Biogeochemical Land-Climate Systems

The evidence of the land cover matters and climate systems is on early paleoclimate model studies and impacts on human-induced deforestation at margined regions. Changes from land conditions due to human activities affect the global climate change. This is driven by changes in emission removals of the GHGs (e.g., CH_4, CO_2, N_2O) by biogeochemical effects. Any land redistribution and local changes on the water vapor and energy between the land and atmosphere influence the biophysical effects on regional climate. The terrestrial biosphere interacts with oceans through influx of nutrients, carbon cycle, water, and particles. This interaction affects crop yield, frequency heat waves, intensive heat waves, rainfall, and air quality. Land has net sources of 441% of the CH_4 emission between 2006 and 2017. IPCC has established GHG inventories and earth system models (ESM) on deforestation and afforestation as to reduce the anthropogenic CO_2 emission. This combines biophysical and biogeochemical processes to the land-climate system processes. IPCC reports that about one-quarter of 2030 mitigation pledges by countries in initial nationally determined contributions (NDCs) under Paris Agreement is expected to come from land-based mitigation (Jia et al. 2019). Some of the mitigation responses have a response option on technical potential for >3 $GtCO_{2\text{-eq}}$ $year^{-1}$ by 2050 through reduction of the carbon dioxide. These technical aspects include afforestation, waste to energy (bioenergy), wastewater recycle, and carbon capture and storage (carbon dioxide removal-CDR). The estimate includes sustainability and cost consideration with social-economic consideration on the climate changes or non-GHG climate forcing. Rising CO_2 concentration limits stomatal opening of the moisture content in the soil, thus reducing transpiration (land-based water cycle/hydrological on climate change). This increases the aerosol levels, declines surface winds, and increases solar radiation to the ground. The impact of the climate-related extremes on land includes the disruption of the water-energy-food production and supply chain, alteration of the ecosystems, hydrological cycle, surface temperature, atmosphere composition, morbidity and mortality, and damages of the infrastructure and settlements and human health. Advanced knowledge and optimization of the mitigation and adaptation with coordination of the sustainable land use and management across all sectors are required to achieve better livelihood, food security, energy security, and water security and improve human health, biodiversity, quality of the local environment, and equitable sustainable development. This is through the modification of the regional and global climate change, seasonal and annual climate variation, extreme weather, and human activities on the land, e.g., deforestation, afforestation, and forest management. Incorporating land-climate processes into climate projection allows emotional intelligence and artificial intelligence to take shape through understanding land's response to climate action and better quantify the potential of land-based response options for the mitigation of climate change (Jia et al. 2019).

Nexus Contributions to Job Market

When developing or implementing a nexus, approaches with regard to minimization of high transaction costs should be an added value in economic measurement. Furthermore, energy and water generation and distribution have high characteristics of being monopolized in the developing countries. This creates merits and demerits to the end users. In general, the CLEWS' nexus has great potential to create new job opportunities and improving living standards in the developing countries. There is a higher need to accelerate the process of the WEF management of emergence of new employment opportunities. Investment of the research and development and implementations on nexus approaches could induce a positive economic effect through job creation. Other disrupted sectors should further be redeveloped, attracting additional meaningful investments and producing new employments within a nexus framework of policies and governance, auditing, and monitoring. Governments have the biggest role to play in the nexus implementation either by making relevant policies or providing funding that subsidizes new technologies and contributes to the welfare of society and enrichment of the emerging market gap.

Carbon Tax

Developing countries' economic transformation and creation of employment through green economy and digitalization will be one of the COP26 green deal agendas. This is with the transition of the national development goals (NDG) 2030 and transition of the low-carbon economy and sustainable development goals as per the economic partnership agreement between the European Union (EU) and developing countries. The commitment through policy-making is paramount in the carbon taxation. The carbon tax is levied on carbon content of fuels (energy and transport sector) and carbon emission trading in the form of carbon pricing (CO_2 equivalent tax/pollution tax). The tax offers potentially cost-effective means of the reduction of the greenhouse gas emissions by shifting the cost from society to companies that create the emissions (the more the emission, the more the cost). South Africa has spearheaded the developing countries in the carbon tax 2019 (carbon tax act No.15 of 2019) meant to curb the carbon emissions (green house emission, industrial processes, fuel combustion, and fugitive emissions) as mitigation to climate change, which is cost-effective and economically efficient. The cost of the impact of climate change to food pricing, water, infrastructure, health, conflict, and disasters tends to shift to polluters (global economy shifts toward lower carbon economy). The impact of crop disease and rainfall changes, extreme storms and drought, and temperature changes has shift changes in food prices and food security. This has direct impact on the carbon taxation (Republic of South Africa ZA 2019).

Data-Driven Nexus

It is a big challenge to relatively obtain national data for the nexus. Small numbers of bodies (e.g., IRENA. World Bank, IEA, UN) share production, transformation, and consumption data. This remains a key challenge in the mitigation of climate change. Examples os data include water evaporations rates from hydro-powers, flood controls or irrigation, water requirement in energy production and food production, mixed SI units on the data presentations, pattern use of water with regard to locations. Data gap is even larger if fully life cycle of the technology is considered (Ferroukhi et al. 2015).

Matters concerning the big data in nexus' availability and accessibility should be of high priority in policy-making. Nexus is based on a holistic environmental and economic perspective, which should use reliable, consistent, and comprehensive data. It is also imperative that data source across the nexus sectors is comparable in terms of resolution and accuracy. Efficiency in implementation of the nexus is dependent on high-accuracy economic databases that support sustainability. Research and development, scientific institutions, and stakeholders can initiate the collection of open data, in order to build a sustainable database for analyzing the nexus. This can add value and be generated only through workable partnerships between the private-public sector, NGOs, knowledge institutes, and local and regional stakeholders (Markantonis et al. 2019).

Industry 4.0 in Nexus and Climate Resilient

The nexus pillars of sustainable development need to be explored in the numerical modelling (i.e., economic modelling; kinetic modelling; life cycle assessment (LCA); WEF modelling; multi-criteria decision analysis (MCDA); CLEWS' modelling; resource flow; network analysis; remote sensing, geospatial, and hydrologic models; finance models; climate models; material and energy models; land-use models; institutional analysis; environmental management; indicator models; social science and integration models; system analysis) (Albrecht et al. 2018). The climate change estimation is based on the sparse station coverage, particularly on scrutiny of

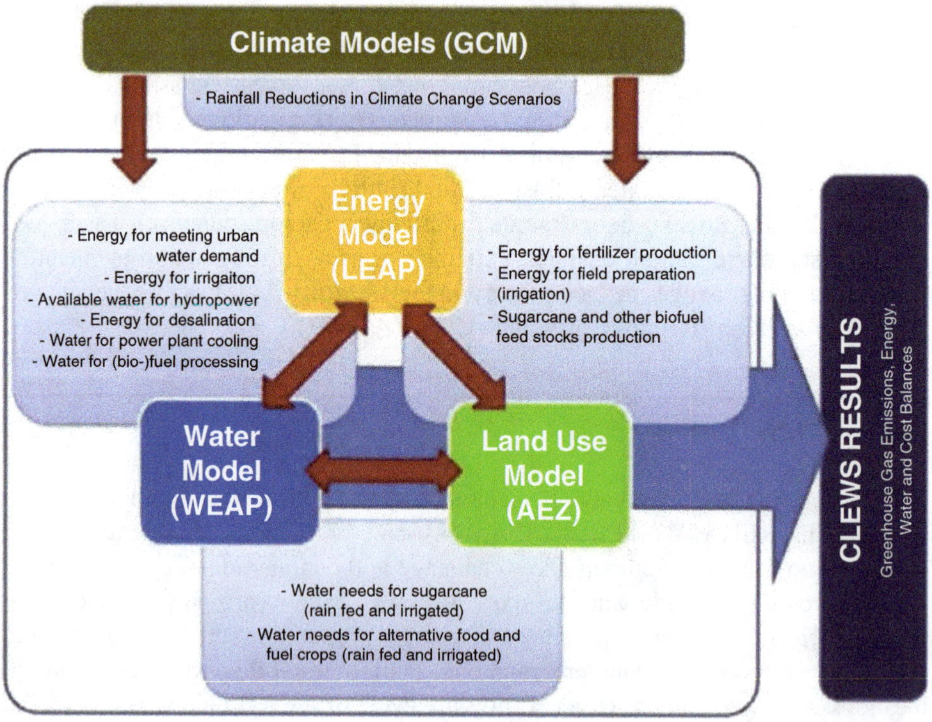

Fig. 4 The CLEWS' framework founded upon the LEAP, WEAP, AEZ, and GCM models

GDP data. Lack of transparency in data sources and collection methods, lack of details on methods of aggression, and lack of metadata have led to differences between GDP estimates, adjustment to historical data, nonrandom errors, and inhomogeneity in time series. Good quality data is paramount to reliable physical and economic modelling of the nexus and climate sector (Conway et al. 2015). Climate, land, energy, and water systems (CLEWS) are integrated. The CLEWS' resource assessment and interlinkages with new paradigms are presented in Fig. 4 (Howells et al. 2013; Welsch et al. 2014).

The CLEWS serve as highlights and dynamics that overlook the system approach to sustainable development. The CLEWS' framework is developed by integrating the LEAP (energy), WEAP (water), GCM (general circulation model, climate), and AEZ (agroecological zoning, land) to produce the GHG emissions, water, and energy balance. The model outcome exercises policy development and resource assessment approach through linking existing single-resource modelling tools (Howells et al. 2013; World Bank Group (WBG) 2016).

Modelling of the Water-Energy-Food-Land-Climate Nexus

The climate-water-energy-food-land are indirectly and directly interlinking. Nexus unique assessment and monitoring is based on indicators and interlinkages to better

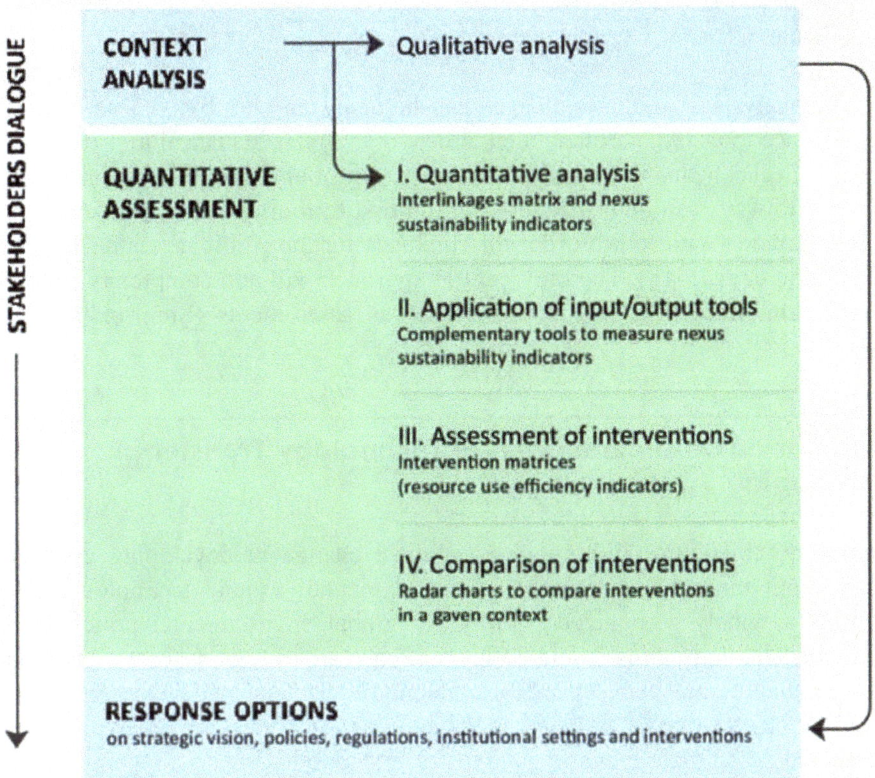

Fig. 5 FAO component of the nexus assessment

understand the potential synergies and trade-offs. Quantification and characterization are mostly needed in decision-making. Nexus modelling and framework are classified as (Hassan et al. 2018; Bizikova et al. 2013):

- Type of models: quantitative analysis models, simulation models, and integrated models
- Geographical scale in addressing intercountry levels

Figure 5 indicates the components of the nexus assessment (Hassan et al. 2018).

For example, assuming change in water (ΔW), reduces freshwater availability and create shift in energy (ΔE) with limiting cooling power that leads to load shedding (power rationing) and water scrubber thus increase shift of climate change (ΔC) and land use. The interlinkages are unique and can go to any directions. The 20 interlinkages are summarized below:

- Water: WL, WF, WE, WC
- Energy: WL, EF, EC, EW
- Land: LF. LW, LC, LE

- Food: FE, FE, FL, FC
- Climate: CF, CW, CE, CL

The analysis of the interlinkages can be analyzed well by the use of artificial intelligence tools (e.g., artificial neural network, ANN), statistical tools (e.g., multi-criteria decision analysis (MCDA)), and conventional modelling concepts. The nexus pathways can be restructured in the first to fourth order. Similar nexus can be constructed with fourth-order interlinkages for the whole nexus making it 120 (Laspidou et al. 2018). The multicentric approach will add complexity with interconnection, trade-offs, and drives on nexus assessments (Simpson and Jewitt 2019).

Adaptation of Climate Change Technology Transfer in Developing Countries

Currently, combating anthropogenic climate change in developing countries is carried out through the transfer of environmentally sound technologies (EST). This is a widely recognized priority for global environmental policy. Climate change technology transfer and policy analytics are of high success factors underlying collaboration between private and public sectors in developing countries. Climate change future policies in developing countries are likely to both be promoted by climate technology transfer and public-private cooperation (cross-sector partnership) through the technology mechanism of the water-energy-food nexus and inclusion of the gender. Building partnership mechanisms can reduce cost and increase local clean development mechanisms (CDM). The difficulties in achieving the CDM include first a sidelined investment by investors on diversifying a project away from cheap forms of climate change mitigation by increasing cost and broader development outcomes. Second, the technology dividends are uncertain and often reflect on preference of host government or deliberative processes involving the stakeholders. Third, green climate fund shortage or project investment guidelines combine production cost, human capacity, and innovator multidisciplinary. Overcoming the barrier allows the smart climate-resilient and efficient integrated energy-water-food nexus system in a developing country: Industry 4.0 with climate change friendly investment to proceed quicker and implement the development dividend (Forsyth 2007). The costs of climate technology transfer and increase local representation in establishing the development dividend can be reduced by cross-sector partnerships (CSPs). CSPs can comprise of orthodox public-private partnerships, where governments make contracts with a private-sector company in order to provide Intelligent Network Infrastructure or services more efficient than the state (Forsyth 2005, 2007). Climate change policy and technology transfer minimize the transaction costs, strengthen the collaboration, build capacity, increase public trust and accountability, and enhance environmental governance under the climate change convention and application of the CDM (Forsyth 2005; COP25 2019b).

Nexus and Research Gap

The concept of the nexus has been received in the business, academics, and policy sectors with a few implemented projects on the ground. The major discussion is how to implement and shift from the ideal concept of the theory based on practice and policies. The nexus methodology emphasis on water-food is majorly researched on hydrological, ecological, and agronomic integrated models and analysis with limited focus on the governance issues of the resources. Versatile methodology to quantify the interlinkages between the WEF and climate change is required. Improving the formulating existing methodology and practices is of high priority to advancing the adoption of the nexus approach. Lack of the reliable data (big data/open source) in the digitalization, electrification, decentralization, and decarbonation world (fourth industrial revolution, Industry 4.0) is a major barrier toward implementation of the WEF nexus approach. Lack of the reliable comprehensive analytic tools is another highlighted concern. Development of critical soft skills (sciences/social/business, STEM) in integrated software and online platform is helpful in addressing the potential synergies and trade-offs in the nexus. Convectional models and AI-based models' development is of a big challenge, and a big question is how nexus and climate change interact and are quantified. In providing important strategies for the SDGs, the integrated models and managements for developing countries is enhanced.

Climate change on the WEF is another critical aspect of high importance on resource availability, distribution, and interconnection to the WEF. How does climate change affect the nexus? What are the impacts of climate change to the nexus? Which agent of change needs to be implemented in WEF NEXUS? Is there political good will in the addressing climate change? What are the governance and policy coordination in place? These fundamental aspects can be addressed based on smart climate-resilient and efficient climate change adaptation, international commitment, corporation and stewardship, new and refurbished green infrastructure projects, green funding, reward and awarding the environmental champions, and payment for ecosystem services (Hassan et al. 2018). The methodological challenge, supports, and opportunities that are associated with robust quantification of the WEF nexus are indicated in Fig. 6 (Chang et al. 2016):

Analytics Framework

The nexus provides a set of guidelines that aims in creating a great and equal level playing field for all sectors while achieving the aims of the SDGs that are multilevel approach and cross-sectoral to climate change. This comprises of six categories (Hoff et al. 2019):

- Nexus framing creates specific understanding on key issues from that which explores the interlinkages between the different sectors and resources.

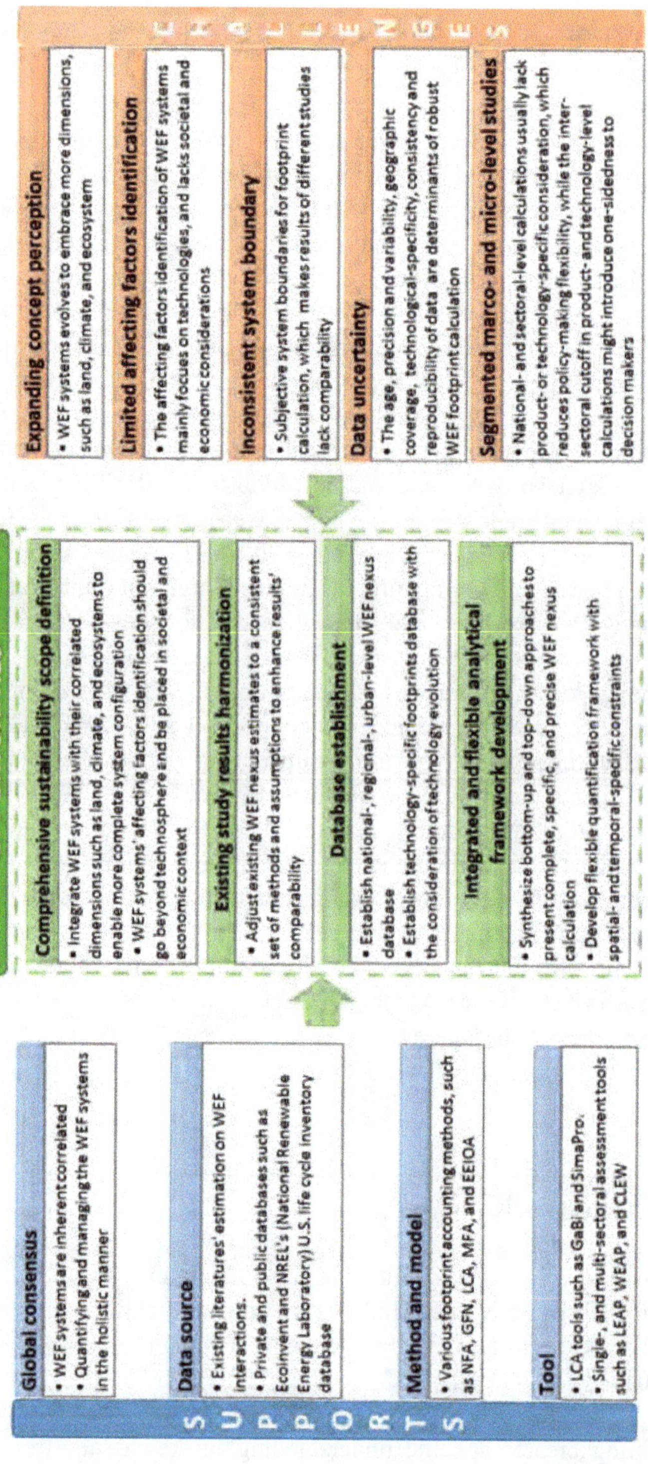

Fig. 6 Methodological challenge, supports, and opportunities that are associated with robust quantification of the WEF nexus

- Nexus opportunities create a category on how to identify a new approach and adding values in context by improving cross resources productivity, reducing resources and environmental degradation, reducing human insecurities/unemployment/poverty, and increasing climate resilience.
- Technical and economic nexus (nexus savings) assesses possible potential benefits from implementation of nexus approaches through multifunctioning production systems, cross resources, and cross-sector recycling.
- Stakeholders' involvement specifies different types and levels of stakeholders involved in the nexus, e.g., private-public sectors and civil society in overseeing the implementations.
- Conditions' framework addresses factors such as policies, technical solution, scalability, initialization, bridging mechanisms, integration of SDGs and NDGs, and innovations (start-ups, incubation, and entrepreneurship).
- Monitoring and evaluation (M&E) serves as an indicator for the required data for implementation of the nexus. This is because it is dynamic objectives, composition of stakeholders, and processes.

The overall methodology of the WEF nexus management approach is performed in three steps (see Fig. 7): (1) overview characterization (identifying and quantifying the connectivity between nexuses), (2) integrated models and analysis (climate-land-energy-water) of the nexus system, and (3) performing future management scenarios to help policy and decision-making (Tashtoush et al. 2012).

The nexus framework is seconded by the holistic resources planning showed in Fig. 8 (Kulat et al. 2019).

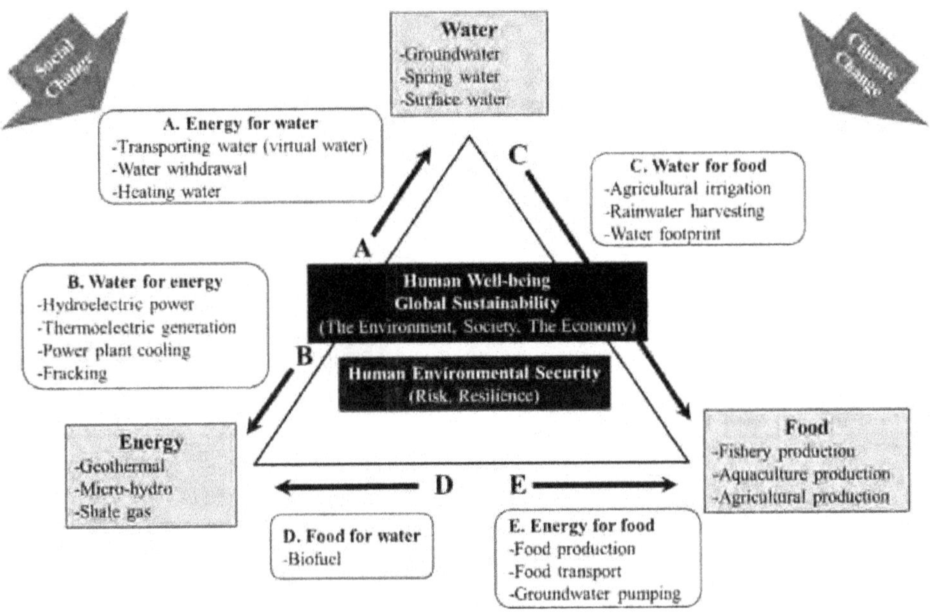

Fig. 7 Water-energy-food nexus analytics framework

Nexus Securities and Environmental Protection

Climate change intensifies high competition and trade-offs between the WEF resources. The potential impact of the climate change is manifested by rise in temperature, more energy demand, and more water demand. Urbanization and population increase (industrialization) account for the 75% of the energy consumption and emission of the 75% of the greenhouse gases (GHGs). This creates high opportunities for the decentralization, decarbonization, and digitalization systems that improve the resource efficiency and implementation of the nexus approached. Suitable adaptation of the climate change demands efficient use of water, energy, and food resources to fight against vulnerability. Figure 9 illustrates the conceptual

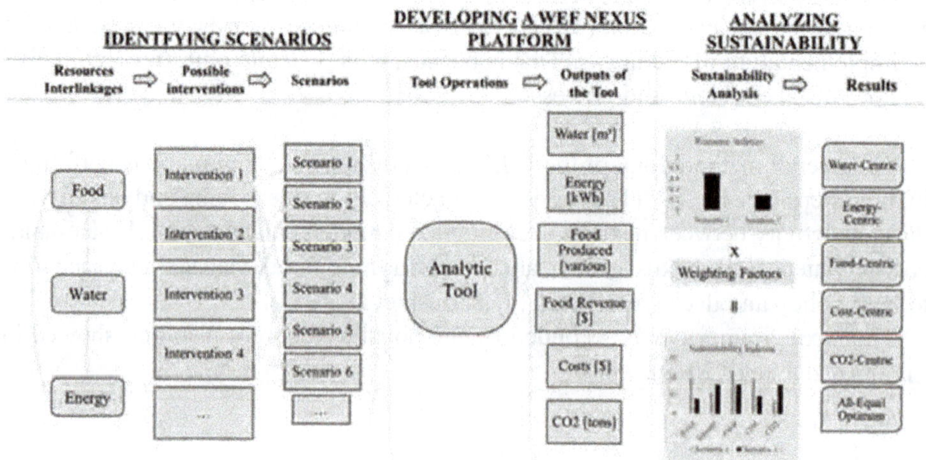

Fig. 8 Holistic water-energy-food nexus resources planning

Fig. 9 Conceptual framework of the water-food-energy nexus and climate change integration

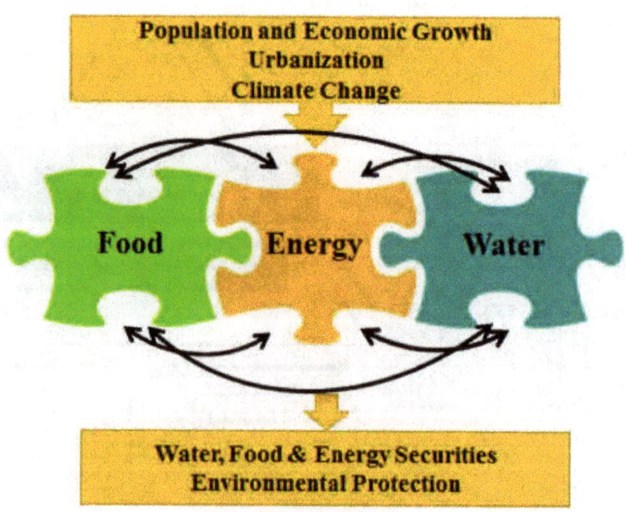

diagram of the WEF nexus with link between the WEF and climate change integration (Tashtoush et al. 2012).

The global emission of carbon dioxide has increased with almost 50% since 1990. The SDGs and NDGs cannot be achieved without addressing the climate emergence (SDGs, UN 2019). The region that is vulnerable to climate change encourages to adapt measures that are implementable and have surety of the WEF security. Enhancing efficiency of small-scale water supply systems and consolidated small services helps reduce vulnerability to climate change of the systems (policy initiative to adapt the climate change in water sector); collaboration in data collection and analysis to mitigation of climate change are key to sustainability of the WEF nexus. Adaptation of the climate change can be improved by building ecological networks that connect ecological to local governments, thus improving decision-making. Spatial equilibrium models seek to optimize resource allocation in various scales, thus providing realistic estimates of the value and into accounting alternative use of the disrupting technology. WEF nexus needs to be integrated to a smart climate resilient with inclusion of the governance and stakeholders. This serves as a comprehensive tool in decision-making toward sustainable resources in climate change (Volk 2014; Daher and Mohtar 2015). Recognition of spatial and sector interdependencies should inform policies, investment, and institutions for enhancing WEF security and climate change (Conway et al. 2015).

Policy and Governance (Coordination and Collaboration) for Climate Change

Climate gas (GHG) emissions and climate risks in countries are not only city government's concerns but also rural government oversight concerns (climate governance). The drivers, consequences, and dynamics of climate change cut across jurisdictional boundaries that require collaboration and coordination of governance across nongovernment and government sectors. Climate change governance is within broader political and social-economic context negotiable with Conference of the Parties (COP) to the UNFCCC (Romero-Lankao et al. 2018). Strategies, policies, and plans provide guidelines and framework for practical coordination and collaboration on CLEWS. Overarching climate change policies integrate the practical collaboration and giving timeline to the policies and their inclusion of climate changes and nexus. Persistence of these strategies recognized a gap in governance (policies and planning). The emphases on a formal model of contract may disincentivize regular collaboration and not only the barrier to cross-sector collaboration and coordination but budge and financial agreements. The issues of the regional security (power imbalance) are a sector interest overriding the climate change agenda. The integrations of the national-level strategies, planning, policies, and age and gender inclusion encourage action on climate change (Pardoe et al. 2018). Increase policy synergies among transformative investments that require a holistic approach with process innovation that fueled a higher level of nonhuman and human capital. Formation of the transboundary with globalized decision-makers

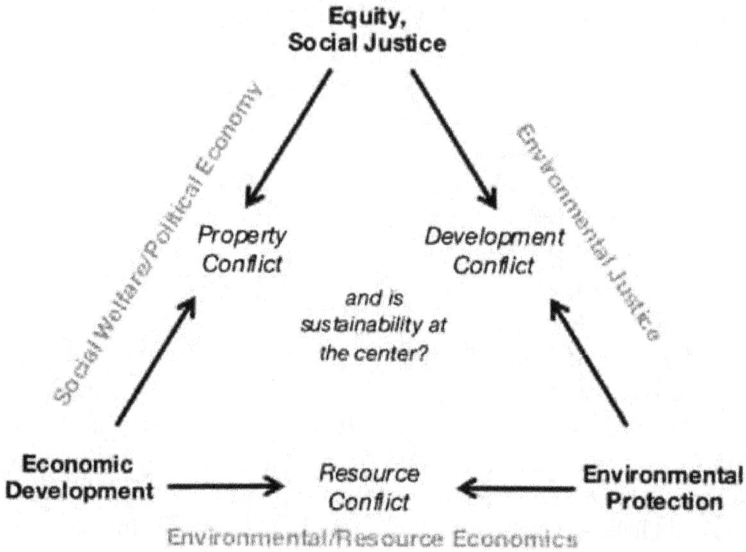

Fig. 10 Three fundamentals of planning

creates integration and enhances benefits through the climate-land and WEF nexus and equitable distribution. Creation and promotion of the gender equality environment enhances the nexus investments and implementations with enhanced family wealth that widens skills and knowledge (Mishra et al. 2019; RES4 Africa Foundation 2018). Collaborative, informed decision making and equitable needs in transformation response of the climate change as well as fundamentals of the legal framework, leadership, land-use regime, land ownership, energy use, food production and distribution, public participation, information sharing, financial resources, growth, and ethos shape the climate change governance (Romero-Lankao et al. 2018). The three fundamentals of planning are presented in Fig. 9. It consists of the development conflict, resources, and property with three broad political/social institutions to manage environmental justice, regulation, environmental economics, and social welfare (Campbell 2013; Fig. 10).

Despite deficiency, state, government, provinces, municipalities, counties, and nongovernmental and governmental action on addressing climate change the jurisdiction over many dimensions of adaptation and mitigations reside on financial and technical capability and policies that hold greatest potential to create legitimate and effective response strategies (Romero-Lankao et al. 2018).

Equity, Climate Change, and Environmental Justice

Major cities are characterized by socioeconomics of different diversifications that are often accompanied with the stratification on gender, race, professional classes, cultures, age, ethnicity, and ability. This gives rise to endure climate stresses and minimize climate risks. Lower social class, disadvantages in the community, hazard prone areas, ethnics, and racial minorities increase their susceptibility to the impact of climate change and reduce their capacity to adapt (Rosenzweig et al. 2015).

Climate change vulnerability is driven by (1) resource influence by social characteristics, (2) ineffective planning and absence of community engagement due to governance and institutional weaknesses, (3) failure of the urban developers in providing accessibility to critical infrastructures and services, and (4) occupational health that leads to physical exposure at different levels. As the climate event such as drought in prone areas becomes more frequent and intense, this can increase the scale and depth of poverty overall. Gender inequality is pervasive and cuts across to disability, income, and literacy that in turn contribute to differential consequences of climate change. Mobilization of the resources increases environmental justice, and equity of climate change requires participation of the locals, international bodies in involvement of the nontraditional resources, finance civil society, public-private sector, monitoring and evaluation, and principles of transparency (Reckien et al. 2018). Figure 11 indicates the interconnected benefits and dimensions of inclusive climate action.

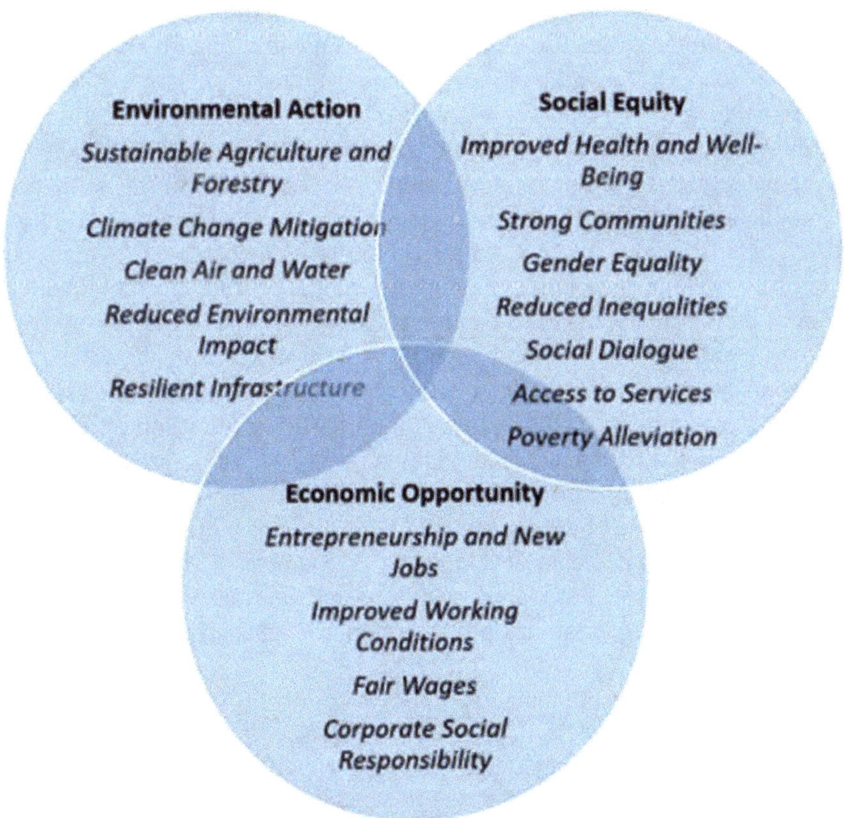

Fig. 11 Interconnected benefits and dimensions of inclusive climate action (Reckien et al. 2018)

The primary goal in achieving environmental justice and equity in climate change is for the urban climate policies to foster human beings, economic and sustainable social development, social capital, access to land, security tenure, risk reduction infrastructure, access to CLEWS; integrate community and stakeholders, in decision-making; adjust climate change policies by ensuring resilience and equity goals; and develop a periodic monitoring and evaluation using fair indicators and progress measurement, resilience objective and feasibility, equitable resources, budgetary transparency, resource allocation scheme, and equitable resilience outcomes (Rosenzweig et al. 2015; Reckien et al. 2018).

Conclusion and Recommendations

The CLEWS nexus is a concept of integration that accesses the resources such as water, food, and energy that are inextricably dependent and linked to one another and have a strong interaction with the environment and climate change. The interlinkages exist between nexuses that manifest in providing exploring opportunities that explore the disruption, policies, and alternatives for increasing the use of resources and sustainable resources management and minimizing the environmental impacts. The nexus and climate resilience offer great opportunities that integrate and manage the disruptive technology in a more sustainable manner by promoting local and regional cooperation and peace; harmonizing the policies, legislation, and strategies; creating proper resource coordination; improving resilience; and reducing vulnerability to attainment of the region integration in-line with SDGs. The impact of the climate-related extreme on land includes the disruption of the water-energy-food production and supply chain, alteration of the ecosystems, morbidity and mortality, and damages of the infrastructure, settlements, and human health. Advanced knowledge and optimization of the mitigation and adaptation with coordination of the sustainable land use and management across all sectors are required to achieve better livelihood, food security, energy security, and water security improve human health, biodiversity, quality of the local environment, and equitable sustainable development. Climate action future policy in developing countries is likely to be promoted by climate technology transfer and public-private cooperation (cross-sector partnership) through the technology mechanism of the water-energy-food nexus and inclusion of the gender. Building partnership mechanisms can reduce cost and increase local clean development mechanisms (CDM). Agenda 2030 sustainable development, Paris agreement, and national development policies are alignments to advance climate-resilient development and green deal that enhance stainable future.

Recommendations arising include:

1. Great investments into nexus with strong collaboration of the multidisciplinary.
2. Governments should work together to achieve green circular economy with smart and intelligent novel solutions.
3. SDGs 17 need to be widely explored and policies clarified.

4. Development of new tools to investigate the interconnection of the WEF nexus with land and climate change.
5. Research and development should be enhanced before decision-makings on sustainability and efficiency.
6. Better understanding is required of the trade-offs between nexus and ecosystems.
7. It is important to take action of the spatial reach of intersectoral interdependencies (pricing, control instruments, and command) and effects in the multi-level governance systems to ensure appropriate participation by affected sectors and stakeholders.
8. Data shortages should be reduced by encouraging open-source digitalized systems with capacity building support on sustainable monitoring and data management systems.
9. Introduction and enhancement of the existing free trade movement on the borders will enhance reaching the SGDs more than expected.
10. Use of the artificial intelligence solutions (nexus and climate change digitalization) on WEF and land and climate change applicability in decision support management, i.e., machine/deep learning, blockchain, remote sensing, IoT, big data, etc.
11. Introduction of the climate change coding (code for climate change and nexus) will enhance early data analytics and action (action toward digitalization, digital earth).
12. Enhanced, efficiency and accessibility of the green climate funding will enable preparation, implementation, and strategic workstreams of national determined contributors and adaptation-related elements of the Paris Agreement. The mechanism will minimize and address loss and damages associated with climate change impacts of developing countries' parties.

Acknowledgments The authors wish to acknowledge the South Africa Water Research Commission (WRC), Project No. K5/2563 for funding; University of Johannesburg (UJ); Enel Foundation; European Union Institute; Process, Energy, Environmental Technology Station (PEETS) funded by Technology Innovation Agency (TIA) through the Department of Science and Technology (DST); SANEDI; City of Johannesburg (CoJ); Gladtech International, Kenya; South Africa's wastewater treatment plants; Innovation Hub; and water utilities, managements, and staffs for capacity building and knowledge transfer.

References

Albrecht TR, Crootof A, Scott CA (2018) The water-energy-food Nexus: a systematic review of methods for nexus assessment. Environ Res Lett 13(4):043002

Birol F (2012) World energy outlook 2012. International Energy Agency, Paris, p 633

Bizikova L, Roy D, Swanson D, Venema HD, McCandless M (2013) The water-energy-food security nexus: towards a practical planning and decision-support framework for landscape investment and risk management. International Institute for Sustainable Development, Winnipeg

C40 (2020) https://www.c40.org/cities. Accessed June 2020

Campbell SD (2013) Sustainable development and social justice: conflicting urgencies and the search for common ground in urban and regional planning. Mich J Sustain 1

Carter S, Gulati M (2014) Climate change, the food energy water Nexus and food security in South Africa. Understanding the Food Energy Water Nexus, WWF-SA

Chang Y, Li G, Yao Y, Zhang L, Yu C (2016) Quantifying the water-energy-food nexus: current status and trends. Energies 9(2):65

Chirisa I, Bandauko E (2015) African cities and the water-food-climate-energy nexus: an agenda for sustainability and resilience at a local level. In: Urban forum 26(4):391–404. Springer, Netherlands

Conway D, Van Garderen EA, Deryng D, Dorling S, Krueger T, Landman W, Lankford B, Lebek K, Osborn T, Ringler C (2015) Climate and southern Africa's water–energy–food nexus. Nat Clim Chang 5(9):837–846

COP25 (2019a) Enhanced Lima work programme on gender and its gender action plan. https://unfccc.int/resource/cop25/cop25_auv_13gender.pdf. Accessed June 2020

COP25 (2019b) Enhancing climate technology development and transfer through the technology mechanism. https://unfccc.int/resource/cop25/cop25_auv_9_TEC_CTCN.pdf. Accessed June 2020

Daher BT, Mohtar RH (2015) Water–energy–food (WEF) Nexus tool 2.0: guiding integrative resource planning and decision-making. Water Int 40(5–6):748–771

Ferroukhi R, Nagpal D, Lopez-Peña A, Hodges T, Mohtar RH, Daher BT, Mohtar S, Keulertz M (2015) Renewable energy in the water, energy & food nexus. IRENA, Abu Dhabi

Forsyth T (2005) Enhancing climate technology transfer through greater public–private cooperation: lessons from Thailand and the Philippines. In: Natural resources forum. Wiley Online Library

Forsyth T (2007) Promoting the "development dividend" of climate technology transfer: can cross-sector partnerships help? World Dev 35(10):1684–1698

Hassan TA, Muhammad K, Ian M, Lars R, Justyna S (2018) The water-energy-food security Nexus; a review on Nexus literature and ongoing Nexus initiatives and policy-makers. Nexus Regional Dialogue Programme (NRD)

Hoff H, Alrahaife SF, El Hajj R, Lohr K, Mengoub FE, Farajalla N, Fritzsche K, Jobbins G, Özerol G, Schultz R (2019) A Nexus approach for the MENA region – from concept to knowledge to action. Front Environ Sci 7:48

Howells M, Hermann S, Welsch M, Bazilian M, Segerström R, Alfstad T, Gielen D, Rogner H, Fischer G, Van Velthuizen H (2013) Integrated analysis of climate change, land-use, energy and water strategies. Nat Clim Chang 3(7):621–626

Jia G, Shevliakova E, Artaxo P, De Noblet-Ducoudré N, Houghton R, House R, Kitajima K, Lennard C, Popp A, Sirin A, Sukumar R, Verchot L (2019) Land–climate interactions. In: Climate change and land. IPCC special report on climate change, desertification, land degradation, sustainable land management, food security, and greenhouse gas fluxes in terrestrial ecosystems

Kulat MI, Mohtar RH, Olivera F (2019) Holistic water-energy-food Nexus for guiding water resources planning: Matagorda County, Texas case. Front Environ Sci 7:3

Laspidou CS, Kofinas DT, Mellios NK, Witmer M (2018) Modelling the water-energy-food-land use-climate Nexus: the Nexus tree approach. In: Multidisciplinary digital publishing institute proceedings

Markantonis V, Arnaud R, Karabulut A, El Hajj R, Altinbilek D, Awad I, Brugemann A, Vangelis C, Mysiak J, Lamaddalena N (2019) Can the implementation of the water-energy-food Nexus support economic growth in the Mediterranean region? The current status and the way forward. Front Environ Sci 7:84

Matheri AN (2016) Mathematical modelling for biogas production. University of Johannesburg, Johannesburg

Matheri AN (2020) Mathematical modelling of biological wastewater treatment and bioenergy production processes. PhD Thesis, Chemical Engineering, University of Johannesburg

Matheri AN, Ntuli F, Ngila JC, Seodigeng T, Zvinowanda C, Njenga CK (2018a) Quantitative characterization of carbonaceous and lignocellulosic biomass for anaerobic digestion. Renew Sust Energ Rev 92:9–16

Matheri AN, Mbohwa C, Ntuli F, Belaid M, Seodigeng T, Ngila JC, Njenga CK (2018b) Waste to energy bio-digester selection and design model for the organic fraction of municipal solid waste. Renew Sust Energ Rev 82:1113–1121

Matheri AN, Eloko NS, Ntuli F, Ngila JC (2020) Influence of pyrolyzed sludge use as an adsorbent in removal of selected trace metals from wastewater treatment. Case Stud Chem Environ Eng

Mishra A, Appadurai AN, Choudhury D, Regmi BR, Kelkar U, Alam M, Chaudhary P, Mu SS, Ahmed AU, Lotia H (2019) Adaptation to climate change in the Hindu Kush Himalaya: stronger action urgently needed. In: The Hindu Kush Himalaya assessment. Springer, Cham, pp 457–490

Newell JP, Goldstein B, Foster A (2019) A 40-year review of food–energy–water nexus literature and its application to the urban scale. Environ Res Lett 14(7):073003

Nhamo L, Ndlela B, Nhemachena C, Mabhaudhi T, Mpandeli S, Matchaya G (2018) The water-energy-food nexus: climate risks and opportunities in southern Africa. Water 10(5):567

Pardoe J, Conway D, Namaganda E, Vincent K, Dougill AJ, Kashaigili JJ (2018) Climate change and the water–energy–food nexus: insights from policy and practice in Tanzania. Clim Pol 18 (7):863–877

Reckien D, Lwasa S, Satterthwaite D, McEvoy D, Creutzig F, Montgomery M, Schensul D, Balk D, Khan I (2018) Equity, environmental justice, and urban climate change: climate change and cities: second assessment report of the urban climate change research network. Cambridge University Press, New York, pp 173–224

Republic of South Africa ZA (2019) Carbon tax act 15 of 2019. https://www.gov.za/sites/default/files/gcis_document/201905/4248323-5act15of2019carbontaxact.pdf. Accessed June 2020

RES4 Africa Foundation (2018) Applying the water-energy-food nexus approach to catalyse transformational change in Africa. RES4Africa Foundation

Romero-Lankao P, Burch S, Hughes S, Auty K, Aylett A, Krellenberg K, Nakano R, Simon D, Ziervogel G (2018) Governance and policy; climate change and cities: second assessment report of the urban climate change research network. Cambridge University Press, New York, pp 585–606

Rosenzweig C, Solecki W, Romero-Lankao P, Mehrotra S, Dhakal S, Ali Ibrahim S (2015) The urban climate change research network (UCCRN) second assessment report on climate change and cities (ARC3–2), and the UCCRN hubs. Cambridge University Press, New York

Scott CA, Kurian M, Wescoat JL (2015) The water-energy-food nexus: enhancing adaptive capacity to complex global challenges. In: Governing the nexus. Springer, pp 15–38

Scott CA, Albrecht TR, De Grenade R, Zuniga-Teran A, Varady RG, Thapa B (2018) Water security and the pursuit of food, energy, and earth systems resilience. Water Int 43(8):1055–1074

SDGs, UN (2019) Goal 13: take urgent action to combat climate change and its impacts. https://www.un.org/sustainabledevelopment/climate-change/. Accessed June 2020

Seeliger L, De Clercq WP, Hoffmann W, Cullis JDS, Horn AM, De Witt M (2018) Applying the water-energy-food nexus to farm profitability in the middle Breede catchment, South Africa. S Afr J Sci 114(11–12):1–10

Simpson GB, Jewitt GPW (2019) The development of the water-energy-food nexus as a framework for achieving resource security: a review. Front Environ Sci 7:8

Stocker TF, Qin D, Plattner M, Tignor SK, Boschung J, Nauels A, Xia Y, Bex V, Midgley PM (2013) Summary for policymakers. Climate change, the physical science basis. Contribution of working group I to the fifth assessment report of the intergovernmental panel on climate change. Cambridge University Press, Cambridge, UK/New York. IPCC-XXVI/Doc.4

Tashtoush FM, Al-Zubari WK, Shah A (2012) A review of the water–energy–food nexus measurement and management approach. Int J Energy Water Resour 3:1–14

UN Water (2013) What is water security. Infographic, 8

United Nation (UN) (2015) Transforming our world: the 2030 agenda for sustainable development resolution adopted by the general assembly. A/RES/70/1

United Nation (UN) (2020) Report of the secretary-general on the 2019 climate action summit and the way forward in 2020. https://www.un.org/en/climatechange. Accessed June 2020

Volk M (2014) Ecosystem services and river basin models: international conference. In: "Sustainability in the Water-Energy-Food Nexus" C03 Special Session: Ecosystems and their Services in the Nexus, May 19–20, 2014. Bonn

Welsch M, Hermann S, Howells M, Rogner HH, Young C, Ramma I, Bazilian M, Fischer G, Alfstad T, Gielen D (2014) Adding value with CLEWS–modelling the energy system and its interdependencies for Mauritius. Appl Energy 113:1434–1445

World Bank Group (WBG) (2016) Southern Africa energy-water nexus. Report no: ACS19178, World Bank

3

Climatepreneurship: Adaptation Strategy for Climate Change Impacts on Rural Women Entrepreneurship Development in Nigeria

C. A. O. Akinbami

Contents

C. A. O. Akinbami (✉)
Institute for Entrepreneurship and Development Studies, Obafemi Awolowo University, Ile Ife, Osun State, Nigeria

Abstract

Adequate and proper adaptation strategies to climate change depend largely on activities in the rural sector, which drives national economy through exploitation of natural resources. Consequently, actions in rural areas are essential to successful climate change adaptation. Rural communities are highly dependent upon natural resources that are affected by climate change, thus affecting their food security, livelihoods, health, and physical infrastructure. Women and their livelihood practices are thereby affected negatively, leading to increased poverty level and low income, because they find it difficult to respond adequately to climate change effects. This study examines the past and existing interventions on climate change adaptation strategies in two rural communities in Oyo State, introduces climatepreneurship strategy, and assesses its effectiveness. This is an explorative study, employing qualitative approach to gather information through in-depth interview (IDIs) from 50 farmers, before and after the interventions. Data collected were analyzed using Atlas ti. This is a powerful workbench for qualitative data analysis using coding and annotating activities to generate different thematic issues for discussions and interpretations with networks. Study revealed that communities had previously experienced some interventions. Such had no impact on livelihood practices because steps to successful intervention were not followed. Socio-cultural practices hinder women development. The newly introduced climatepreneurship strategy improved livelihood practices. Study outcomes will expectedly be integrated into policy framework for sustainable rural women entrepreneurship development and also replicated in other rural areas in Nigeria.

Keywords

Climatepreneurship · Adaptation strategies · Climate change · Rural women · Entrepreneurship development · Nigeria

Introduction

In recent years, the terms viable adaptation strategies and climate resilience have become important concepts for both scholars and practitioners working on climate change. This progress reflects a growing interest from various disciplines in a holistic understanding of complex systems, including how societies interact with their environment and adapt to the increasing effects of climate change. This new lens offers an opportunity to focus on communities' ability to prepare for and adapt to the challenges posed by natural hazards and the mechanisms/strategies that have been developed to cope and adapt to threats. This is important because repeated stresses and shocks still cause serious damages to communities across the world, despite efforts to better prepare for disasters. This is especially relevant to the rural communities that are highly dependent upon natural resources which are affected by climate change, thereby affecting their economic activities, food security, livelihoods, health, and physical infrastructure.

Women in particular and their livelihood practices have become highly vulnerable. They face various challenges from effects of climate change such as drought, delayed rainfall pattern, heat wave, flood, etc. According to Howden et al. (2007), agriculture as a major economic, social, and cultural activity of most women, particularly in the rural areas, provides a wide range of ecosystem services. The sector is saddled with the responsibility of meeting the projected growth in human population and per capita food demand. Unfortunately, this sector remains highly sensitive to and affected by climate variations. The economic activities are therefore affected negatively, leading to increased poverty level and low income, because the rural women find it difficult to respond adequately to climate change effects.

Many studies have argued that climate change and its effects are clear, escalating and will continue to be a contending issue around the world, because enormous quantities of greenhouse gases (GHGs) have been released into the atmosphere over the past decades through anthropogenic activities (UNDP 2010; Adesina and Odekunle 2011; IPCC 2014, 2015; Salami 2010). According to Dube and Phiri (2013), climate change is now a well-accepted reality with emerging evidence that climate change poses a massive threat for development especially in poor countries. Also, Sadashivam (2010) reported that climate change is a global and regional environmental problem faced by humanity and livelihoods with strong implications for rural populations, particularly women. Kumar and Sharma (2013) analyzed the impact of climate change on agricultural productivity and submitted that variation in climate affects food grain and non-food grain productivity and other livelihood practices. This is due to the fact that climate change often alters the physical geography of an area leading to a disappearance or reduction of natural habitats that constitute the livelihoods of the women. Declining precipitation and rising temperatures are making farming increasingly more difficult with declining productivity, and thus aggravating food insecurity.

In Nigeria, Nwokeoma et al. (2017), Ikehi et al. (2014), Akinbami et al. (2015, 2019), and Ali (2011), have reported on the effects and social consequences of climate change on the farming families, livelihood practices, farmlands, economic activities, food security, health, and physical infrastructure in various communities in the country. These studies reveal that global warming which is as a result of climate change has serious impacts on the communities and their livelihood practices, and these effects will spiral out of control, if the current trend of increasing greenhouse gas emissions is not curbed. This will bring about more natural disasters, increased declining livelihood practices, worsened global food supply, and also threaten the existence of millions of people, with a disproportionately high impact on the impoverished and indigenous populations, particularly, the women. Therefore, there is a need to continue to pursue climate change mitigation and adaptation strategies with vigor (IPCC 2014).

Climate Adaptation Strategies Among Rural Women in Nigeria

Some studies have documented various adaptation strategies geared toward economic development of women in Nigeria (Ikehi et al. 2014; Abid et al. 2015; Etim and Etim 2019) such as planting date adjustment, crop diversification, conservation agriculture, and crop and livestock integration. However, according to Yilmaz (2014) and Etim and Etim (2019), many interventions have failed due to low awareness level and lack of proper involvement of stakeholders, inappropriate non-local technologies to the areas affected, and improper communication of interventions and implementation as the top-down approach was adopted, without the participation of community members. Howden et al. (2007) also alluded that there are many potential adaptation options available with substantial benefits only under moderate climate change for some cropping systems and are therefore not very effective under more severe climate change. In addition, most interventions lack long-term plan for monitoring and sustainability.

In order to build the resilience and sustainable development of women across the whole disaster cycle, involvement of communities in identifying problems (community communication) and involving them in addressing the identified problems become key factors to be considered in determining information-based adaptation strategies that are borne out of earlier sensitization programs. This is more of a bottom-up rather than top-down approach. It is believed that this approach will reduce the impacts of climate change as well as disaster-related losses and lead to faster recovery and the ability to "build back better" (Cutter et al. 2008).

Joo et al. (2011) further expatiated on steps that will aid sustainable development through implementation of interventions as (a) identifying key performance measure for the intervention by creating a plan to know the community to have an idea of their problem, attitude, and behavior; (b) measuring own's internal performance level by how you have to be knowledgeable about the identified intervention and able to gather a team of experts to handle the identified problem and work on the actual and appropriate intervention; (c) identifying intervention advantages and disadvantages, process, and training models and materials; and (d) implementing intervention to close a performance gap by holding meetings and executing all the planning that have been done and importantly, follow through to achieve successful implementation which is monitored thereafter.

Therefore, in responding to the need of implementing appropriate measures to address the severity of climate-related hazards among Nigerian rural women, adequate and effective adaptation strategies are necessary to handle the effects of climate change. Such strategies will integrate the findings of climate change risk assessments into planning processes for disaster risk reduction and management and embrace methods that are climate smart and able to turn climate change challenges to entrepreneurship opportunities, thereby reducing poverty among the rural women, in particular, hence, the introduction of climatepreneurship.

Climatepreneurship and Other Smart Strategies

In literature, scholars had conducted studies on smart strategies and termed differently, such as agri-preneurship, ecopreneurship, etc. Agri-preneurship, according to Ashe (2019), is a new measure taken by developing economies utilizing various forms of technological tools and methods to tackle farm yield, which is sometimes referred to as smart farming. Pai, Shah, and Bohara (2020) also stated that, as the effects of climate change and population growth resulted in high food demand, smart farming is now being adopted. Agri-preneurship comprises of some farming methods such as geo-mapping, greenhouse farming, hydroponics, nutrient cycling, and soil analysis that could address the issue of low farm yield in developing economies (Ariani et al. 2018).

Ecopreneurship was seen as an answer to market breakdown in dealing with negative environmental impacts caused by the industries by focusing on underlying green values and solving the problems in the society caused by these businesses (Kirkwood and Walton 2010). That is, business activities should be geared toward providing values across three dimensions: economic, social, and environmental. Priority is given to addressing the effect of the negative externalities of firms' economic activities on these firms' immediate surroundings (Doman'ska et al. 2018). The goal is, therefore, to build a business model that is sustainable in the long run which ecopreneurship contributes to achieving this goal by seeking simultaneously for both profit and environmental sustainability (Rekik and Bergeron 2017). The concept of ecopreneurship is based on three pillars: innovation, caring for the environment, and long-term sustainability (Hultman et al. 2016).

Climatepreneurship is an adaptation strategy to climate change effects which consists of activities that are climate smart and make use of locally available resources for economic empowerment and enhancement of the social status of rural women. It also helps to turn climate change challenges into opportunities, and thereby emphasizes livelihoods' diversification. This concept entails climate change, adaptation strategies, and entrepreneurship, while according to Omodanisi et al. (2020), smart agri-preneurship entails smart agribusiness, smart technology, and entrepreneurship. Entrepreneurship in agri-preneurship is defined as the means to create and develop a profitable agribusiness, while in climatepreneurship, it is a means of turning climate change challenges into business ventures in all livelihood practices, as well as developing profitable agribusiness by adopting adaptation strategies. That is, agri preneurship deals with ways of increasing farm yield, while climatepreneurship deals with climate-related issues as they affect the livelihood practices and more importantly, turning the challenges of climate change to various business ventures. On the other hand, ecopreneurship deals with the means of carrying out entrepreneurial activities in order not to affect the environment negatively.

Though, the concepts are geared toward achieving sustainable development and reducing starvation among the populace, climatepreneurship goes a step higher by introducing livelihood diversification, exposure to the use of natural resources in the local environment, and turning climate change-related challenges into business opportunities as a major adaptation strategy (Rehman and Shaikh 2014; FAO 2011). Agri-preneurship can therefore be seen as a subset of climatepreneurship aimed at increasing farm yield and enhancing food security in an eco-friendly manner (FAO 2018; Lin et al. 2015; Omodanisi et al. 2020; Ponisio and Ehrlich 2018; Adebiyi et al. 2018). Succinctly put, climatepreneurship is an adaptation strategy to climate change effects which consists of activities that are climate smart, eco-friendly process, and making use of locally available resources for economic empowerment and enhancement of the social status of rural women. It also helps to turn climate change challenges into opportunities.

Thus, climatepreneurship can be regarded as a viable and significant option to safeguard rural livelihoods and ensure rural sustainable development. Hence, this study on climatepreneurship, which presents viable adaptation strategy for addressing climate change, impacts women entrepreneurship development in Nigeria. It aims to examine the past and existing interventions and assess the effectiveness of implemented climatepreneurship strategy introduced by this study among rural women in selected communities in Oyo State.

Theoretical and Conceptual Framework

Theoretical Framework

This study is anchored on socio-ecological system approach which sees a community as a coherent system of biophysical and social factors that regularly interact in a resilient, sustained manner and are linked systems of people and nature, emphasizing that humans must be seen as a part of, not apart from, nature (Berkes and Folke 1998; Redman et al. 2004). The socio-ecological model (SEM), therefore, provides the framework for understanding the interactions between the multiple levels of a social system and how behavior and attitude are thereby affected. The multiple levels according to SEM are individual, interpersonal, community, organizational, and enabling environment (Fig. 1). The main tenant of this theory is based on

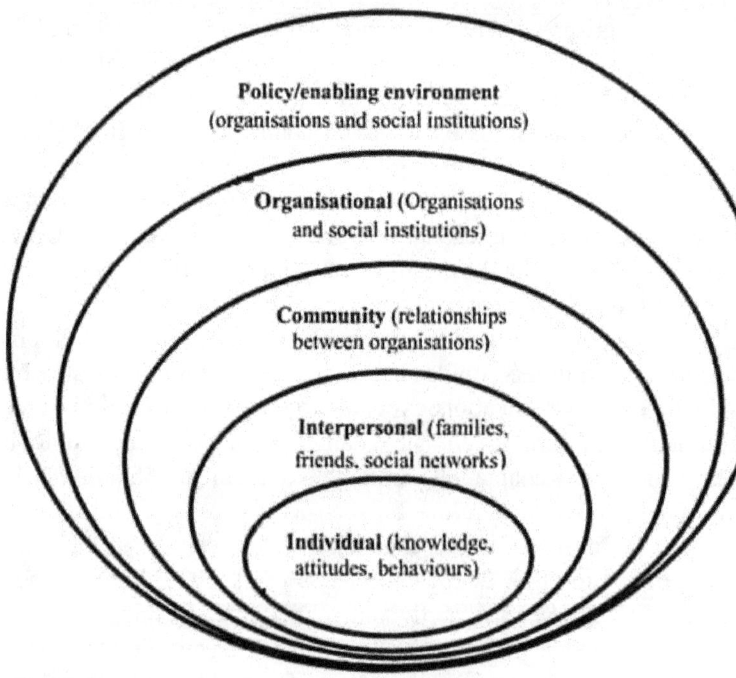

Fig. 1 Socio-ecological theory. (Adapted from Bronfenbrenner (1989))

cooperation between individuals and environment within the system with emphasis on the model of communication for development (C4D), how to incorporate social norms and needs into program planning and partnerships to finally achieve capacity strengthening (CDC 2014).

Consequently, in this regard, adaptation strategies have been acknowledged as useful measurement to understand to what extent the intervention measures address the needs of people as end-users, to what conditions people are still vulnerable, whether the potential outcomes are expected or unwanted (MOVE D.5 2011; Yilmaz 2014). In this sense, it is important to monitor the effects of implemented strategies, especially in terms of observing how the affected communities adapt economically and socially.

Conceptual Framework

The conceptual framework (Fig. 2) further explains the operationalization of the concepts in order to achieve positive results in the rural areas. The individual (rural woman) is the victim of climate change impacts, norms, and socio-cultural activities, and with low livelihood practice, who equally needs to change her own personal attitudes and beliefs toward climate change. Though, aware of the effects of climate change on her livelihoods but yet helpless because of the major and drastic changes

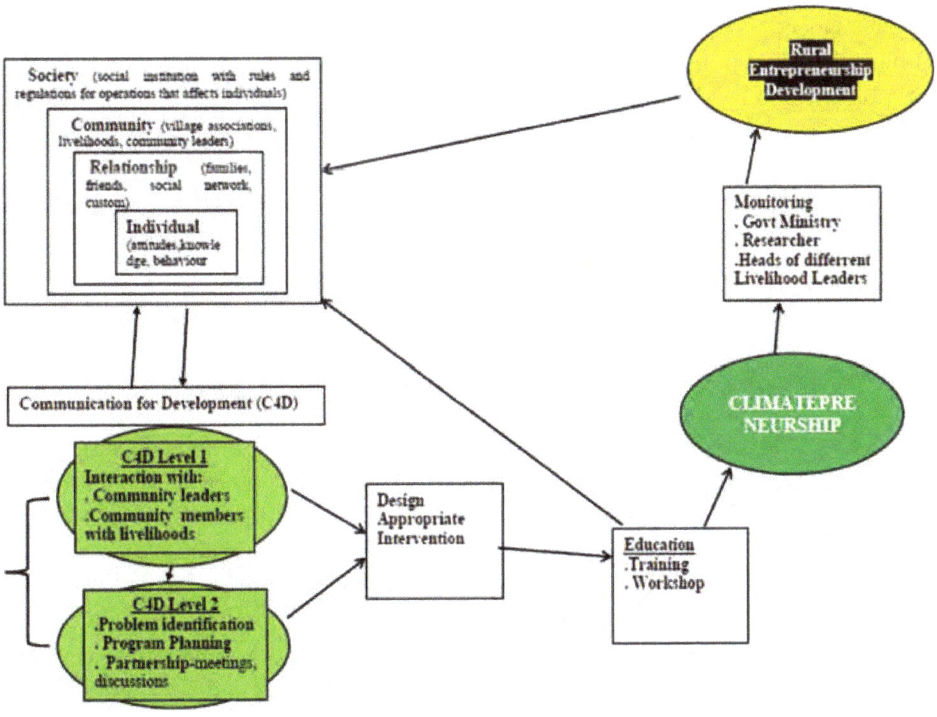

Fig. 2 Conceptual framework. (Source: Author)

in weather being experienced and leading to low farm yields and sales. Even though currently taking steps of engaging in moderate activities in order not to be idle, the poverty level is still on the increase. Having awaited help from government circle for so long, and now frustrated, the victim of the climate change impacts seeks for solution to the economic challenges. The community houses the relationships among institutions and informational networks within defined boundaries, which comprises of the total environment (social, economic, norms, and various association groups) as they affect the individuals reflecting gender specifics. The society at large dictates the enabling environment at the local, state, and national levels including policies that affect and regulate the choices the individual makes.

This framework underscores communication as a vital instrument for development, with the argument that livelihoods have been left to suffer as a result of non-involvement of the community members (i.e., partnership) in most of the interventions/programs that have at various times been designed for rural development. Consequently, the approach to the implementation of such programs had been top-bottom approach, and this has led to the failure of most of the interventions. If interactions through stakeholders meetings take place in order to identify actual problem and gain better understanding before planning a program, clarify limiting socio-cultural issues, designing appropriate training on handling and management of intervention, and establishing a monitoring team (as part of the monitoring and evaluation mechanism), then the designed program will definitely be successful, and the livelihoods of individuals will improve with poverty level reduced and eventually leading to sustainable development in the rural areas.

In essence, in order to reduce disaster and build resilient communities, the socio-ecological model should be an effective learning and assessment tool for the development of rural livelihoods. As the increasing new information and training on how to improve livelihood practices amidst impacts of climate change are gained, the employment of both bottom-top approach and partnership method will stir up a strong will among the rural dwellers to follow through with the new intervention (s) designed. The society and communities will directly be affected positively since both are collections of individuals.

Methodology

This study was carried out to determine the extent to which adopted adaptation strategies to climate change have impacted on the rural women entrepreneurship development in Nigeria. An assessment study (Akinbami 2015) had been carried out to determine the impact of climate change on the livelihood practices of women in the rural areas of some southwest States of Nigeria. Due to limited fund available for the interventions in this study, Oyo State and some of the communities in the State were purposively chosen based on the assessment report on vegetation zone and availability of needed resources (Akinbami 2015) to pilot the implementation of the climatepreneurship strategy. This is an explorative study, employing qualitative approach to gather information through in-depth interview (IDIs) and questionnaire administration from 50 farmers, both male and female, before and after the interventions. This study adopted descriptive research design to examine the past and existing interventions in the last few years before the introduction of climatepreneurship strategy. The study area is in Oyo State which is in the southwest region of Nigeria.

The study was carried out in two purposively selected communities – Araromi Idowu and Abokede Communities at Ido LGA, Ibadan, Oyo State, Nigeria (Fig. 3). The communities are made up of men and women whose different livelihood practices have been affected negatively by climate change. The target populations for this study are mostly women and few men involved in crop farming activities (cassava, maize, oil palm) and who were able to supply land space for planting. Pre-intervention interview was conducted only to participants who were able to supply up to an acre of land each. This was to capture necessary immediate situational experiences of the women and be able to compare with post-intervention interview.

Some men were also added for them not to view the project as just solely pro-women activity and a means of empowering the women against the men. This became useful in order to address the existing socio-cultural issues. The instrument for primary data collection was structured interview questions developed from literature reviewed for the study with inputs from the pre-intervention stakeholders meetings. The use of in-depth interview was to harvest the views and experiences of the respondents on the nature of perceived climate change impacts on their livelihood practices and previous interventions in their communities. The IDIs targeted participants who could provide up to an acre of land to accommodate intervention varieties, more for the purpose of monitoring and evaluation while the questionnaire was distributed among the other participants who showed interest in the intervention.

The project targeted 50 participants who had portions of land available and who were interested in participating.

The employed methodology involved some activities, such as:

- Community visitation and sensitization program.
- Meetings with different stakeholders in the communities to identify challenges.
- Distribution of improved drought-tolerant maize varieties and beta-carotene fortified cassava for planting.
- Distribution of hybrid oil palm trees (those who were able to supply up to an acre of land each) for planting,
- Training of women on entrepreneurship businesses in the oil palm tree products.
- Meetings with government parastatals saddled with women affairs, climate change, and agriculture in the State to intimate them with the objectives of the study and solicit their partnership and collaboration for its success.
- Facts finding workshop for all stakeholders involved in the study.
- Agricultural extension officers from the State Government were also part of the training exercise and workshop for effective monitoring of the intervention.

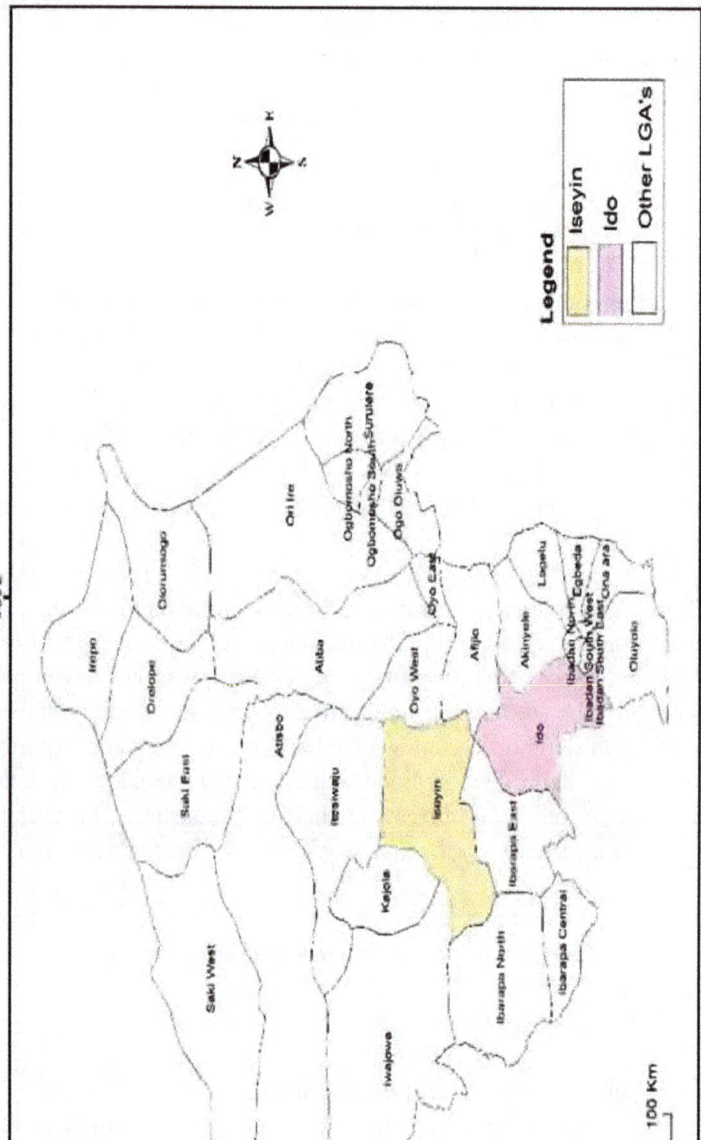

Fig. 3 Map of Oyo State showing the Ido Local Government Area where the study areas are located

Data Analysis

Data collected were analyzed using Atlas ti. This is a powerful workbench for qualitative data analysis using coding and annotating activities to generate different thematic issues for discussions and interpretations with networks.

Result and Discussion

Socio-Demographic Characteristics of Respondents

This section provides descriptive summary of the characteristics of the rural women and men entrepreneurs (18) gathered during IDIs, who were able to supply up to an acre of land each for the intervention in Table 1. However, the other 32 participants who showed interest in the exercise and with less than an acre of land were also accommodated by the intervention. The 18 participants reported in Table 1 were given drought-resistant maize, vitamin A-enhanced cassava, and hybrid oil palm trees. In addition, the intervention provided drought-resistant maize and vitamin A-enhanced cassava to the 32 participants (those with less than 1 acre of land) for planting on their farming space.

All the 18 participants were in the age range of 40–65 years (Table 1). Participants were mainly into crop farming involving planting of maize grains, cassava, and palm tree. All of the women were found to be involved in another occupation, such as food processing of garri (cassava-based Nigerian staple food); cassava and palm oil ventures; and sales of cloths, beverages, and other food items. While all the men had crop farming as their major economic activity, about 67% were not involved in any other economic activity. The table further revealed that about 20% of the men were into additional occupation of commercial transportation using their motor cycles, and it was observed that these are of the younger generation. About 10% of the men were involved in food processing as an additional economic activity.

Table 1 also showed their educational status; about 72% of the participants were not with formal education, 22% had only primary school education, and 6% were with secondary school education. About 96% of the participants had spent more than 15 years in crop farming business.

The age range of the other 32 participants is between 38 and 65 years, with 80% being women involved mostly in food processing, while the men (20%) were mainly into crop farming. It was observed that the educational status, awareness of climate change, and involvement in other occupational activities are the same as those respondents in IDI. The opinions of the respondents in both instruments are presented in the analysis.

Awareness of Climate Change and Impacts on the Business Activities

All the respondents (from IDIs and questionnaire) reported that they are aware of the changes in the weather recently. The participants stated that the changes had been

Table 1 Socio-demographic factors of participants during IDIs

Respondent	Age	Sex	Educational status	Occupation	Community	Years of farming engagement	Second occupation
Respondent 1	40	Female	Secondary	Crop farming	Abokede	5	Sales of provision
Respondent 2	60	Female	No formal education	Crop farming	Abokede	30	Sales of cloth
Respondent 3	50	Female	No formal education	Crop farming	Abokede	25+	Food processing (palm oil)
Respondent 4	40	Female	No formal education	Crop farming	Abokede	20	Food processing (palm oil, cassava)
Respondent 5	65	Female	No formal education	Crop farming	Abokede	30+	Food processing (palm oil)
Respondent 6	50	Female	Primary education	Crop farming	Abokede	20+	Sales of yam flour
Respondent 7	40	Male	Primary education	Crop farming	Araromi-Idowu	15	Commercial motor cycling
Respondent 8	52	Male	Primary education	Crop farming	Abokede	30+	None
Respondent 9	60+	Male	No formal education	Crop farming	Abokede	30+	None
Respondent 10	62	Male	No formal education	Crop farming	Abokede	30+	None
Respondent 11	55	Female	No formal education	Crop farming	Abokede	20	Food processing (palm oil)
Respondent 12	62	Female	No formal education	Crop farming	Araromi-Idowu	30+	Food processing (palm oil)
Respondent 13	42	Male	No formal education	Crop farming	Araromi-Idowu	30+	Commercial motor cycling
Respondent 14	57	Male	No formal education	Crop farming	Abokede	30+	None
Respondent 15	60	Female	No formal education	Crop farming	Araromi-Idowu	30+	Sales of dry maize
Respondent 16	50	Male	No formal education	Crop farming	Abokede	30+	None
Respondent 17	56	Male	No formal education	Crop farming	Abokede	30+	Food processing (palm oil)
Respondent 18	48	Male	Primary education	Crop farming	Araromi-Idowu	30+	None

Source: Study Survey

observed through personal experiences and comments from colleagues as far back as 30 years' ago, and this has been confirmed by information from the radio programs they listened to on climate change. Their awareness cut across the elements of weather vagaries such as distortion of rainfall pattern, drought, sea level rise, temperature rise, and increase in humidity. They reported "as farmers, we have experienced hard times as a result of the changes, which ranges from delayed rainfall to increase in temperature and drought." Consequently, the communities have suffered some economic hardship and social consequences from impacts of climate change. Some of the responses of the respondents are:

> There is too much heat, as a result, our crops, vegetables and other things planted do not do well. Sometimes, the farm products harvested are both too few in quantity and small in size that we do not have good quality crops to sell. This has increased poverty level in this community, especially among the women. **Female Participant, Abokede Community, 40 years old**
>
> The weather has become so unfriendly, we as farmers do not have rains as at when due, the sun is too much, our crops therefore produce low yield. Ultimately, we have little to sell and are having poverty increasing in this community. **Male Participant, Araromi-Idowu Community, 42 years old**

This study outcome supports the work of Suleiman (2014) which stated that farmers had crops uprooted prematurely and destroyed due to harsh weather condition; as such, crops could not be taken to the market. They lost money that should have been gained.

This could be the reason why the younger-generation men in this study subscribed to another business (commercial transportation) because the pangs of impacts of climate change were being felt. The income generated from only farming activity was no longer enough to cater for their needs. These impacts had led to the increase of poverty level in the communities and particularly, among the women. It has further shown the relationship between climate change, livelihood practices, and community empowerment.

Also, due to the low level of education of the rural communities, participants could not process the information received into finding out the causes of climate change and what adjustments are needed to be made to improve the livelihood practices. Participants knew nothing about crop diversification strategy which could have helped to improve the livelihood practices. This study revealed the communication gap that exists between the rural areas and research outcomes. This submission is in line with FMENV (2010) which stated that programs, policies, and activities of the Ministry on climate change do not seem to have specifically targeted and involved farmers. Consequently, the farmers are exposed to hardship. In affirmation, participants from both communities stated:

> We are hoping that help will come one day. Our community leaders know about our plight since we are in it together, unfortunately, they don't know what to do. **Female Participant, Araromi-Idowu Community, 62 years old**
>
> We don't even know the cause of the change nor do we know the solution. **Male Participant, Abokede Community, 57 years old**

Participants could not produce any observed weather pattern based on the reported climate change, to help them know when and what to plant, that is, adjusting the planting calendar in order to adapt to the change in rainfall pattern, as stated by Ikehi et al. (2014) that "planting date adjustment" is one of the strategies that can be employed by the rural communities to cope with impacts of climate change. The study is also of the opinion that the low level of education of the participants may be a major factor responsible for this inability.

Participants declared that they have been coping through prayers. They perceived climate change as an act of God and not induced by the activities of man. They also said God has decided to punish man as a result of sin which is on the increase daily. This has affected their readiness and preparedness for mitigation and adaptation strategies. A participant said:

> Sacrifices were done by our forefathers with animals and now are being done with human beings, so, why will God not be angry. **Male Participant, Araromi-Idowu Community, 62 years old**

The participants' lack of understanding therefore calls for urgent and proper training at the rural areas by the agriculture extension officers, so that rural dwellers especially will not continue waiting for the government to handle what they can quickly put right by themselves. This study sees this step as important because an enlightened community will be able to contribute to the process of developing adequate adaptation strategies which the community will own and preserve. Consequently, intensive and extensive sensitization and education programs mounted by both public and private sectors including the faith and community-based organizations as well as climate-based NGOs are essential since effective adaptation requires a holistic approach that incorporates both policy, people, and investment issues into the planning and decision-making processes. This is of great importance as no one solution will adequately address effective adaptation in all communities (UNDP 2010). Effective adaptation will require long-term planning approaches at the national, regional, and also, local level, which is responsible for the national food supply. However, at all the levels, their associated impacts in the rural communities may not succeed if the end users are not carried along due to the lack of understanding of subject matter.

Experiences of Participants and Their Livelihood Practices on the Effects of Climate Change

The study revealed some of the experiences of participants from the effects of climate change as it affects their livelihood practices. According to the participants, the climatic change ranges from delayed rainfall, strong wind, increase in temperature, and heavy rainfall leading to flooding, thereby affecting the roads and also resulting in overflown rivers. It was observed that participants from the two communities have similar experiences as they commented on the effects of different aspects of climate change. It was reported that there has been delayed occurrence of rainfall, and when it occurs, the intensity was unusually heavy and had an adverse effect on their farm products. This was reported by participants during the in-depth interview sessions and questionnaire administration conducted in the two communities. The reported effects of climate change on the participants' crop farming business are presented in Fig. 4. According to Akinbami et al. (2019) and Slater et al. (2007), these negative impacts have implications on food security of the nation. As a nation, researchers, civil society, and policy makers should be in quest for innovative approaches to food security and not depleting the available quantity, especially with population increase which may expose people to the risk of hunger and its associated social vices.

Previous Intervention in the Two Communities

According to some of the participants from Araromi-Idowu and Abokede communities, it was reported that they have never been part of any intervention either from the government, NGO, or a researcher. The community leader at Araromi-Idowu reported that there has not been specific intervention from the government except for a borehole that was dug for the community about 3 years ago. The borehole has ceased functioning due to lack of electricity to pump constantly. However, he further stated that some researchers in the last 2 years had visited the communities with some interventions. He retorted:

> About two different times, some people (researchers) visited our community and distributed cassava stem (they said it is fortified with vitamin A) and maize grains to few male farmers. We were not taught the importance of the seedling and grains. As a result, most of us did not plant them. I kept the grains in my house and I know that my colleagues did same. **Youth Community Leader, Araromi-Idowu Community**

From the submission of the respondents, it was observed that interventions did not succeed because the measures that lead to successful implementation of interventions were not taken into consideration. Measures such as communication and partnership with the communities' members, as well as monitoring, were missing (Joo et al. 2011; CDC 2014; Yilmaz 2014; Etim and Etim 2019). Apart from the ineffectiveness of the previous interventions, the women reported special challenge they are facing as a result of socio-cultural practices which are still very strong in the rural areas.

> About three years ago, some people came to train us women on how to manage our businesses, the men did not allow us to participate fully. They felt threatened and it is also against the rules/practices of this community to gather only the women. The cause of their anxiety is that they thought we would be exposed to things that will make us rebel against them, especially, when we are financially liberated. **General response from women in the two communities**

This study is of the opinion that these unattended to changes in weather patterns through various appropriate and effective adaptation and mitigation measures will

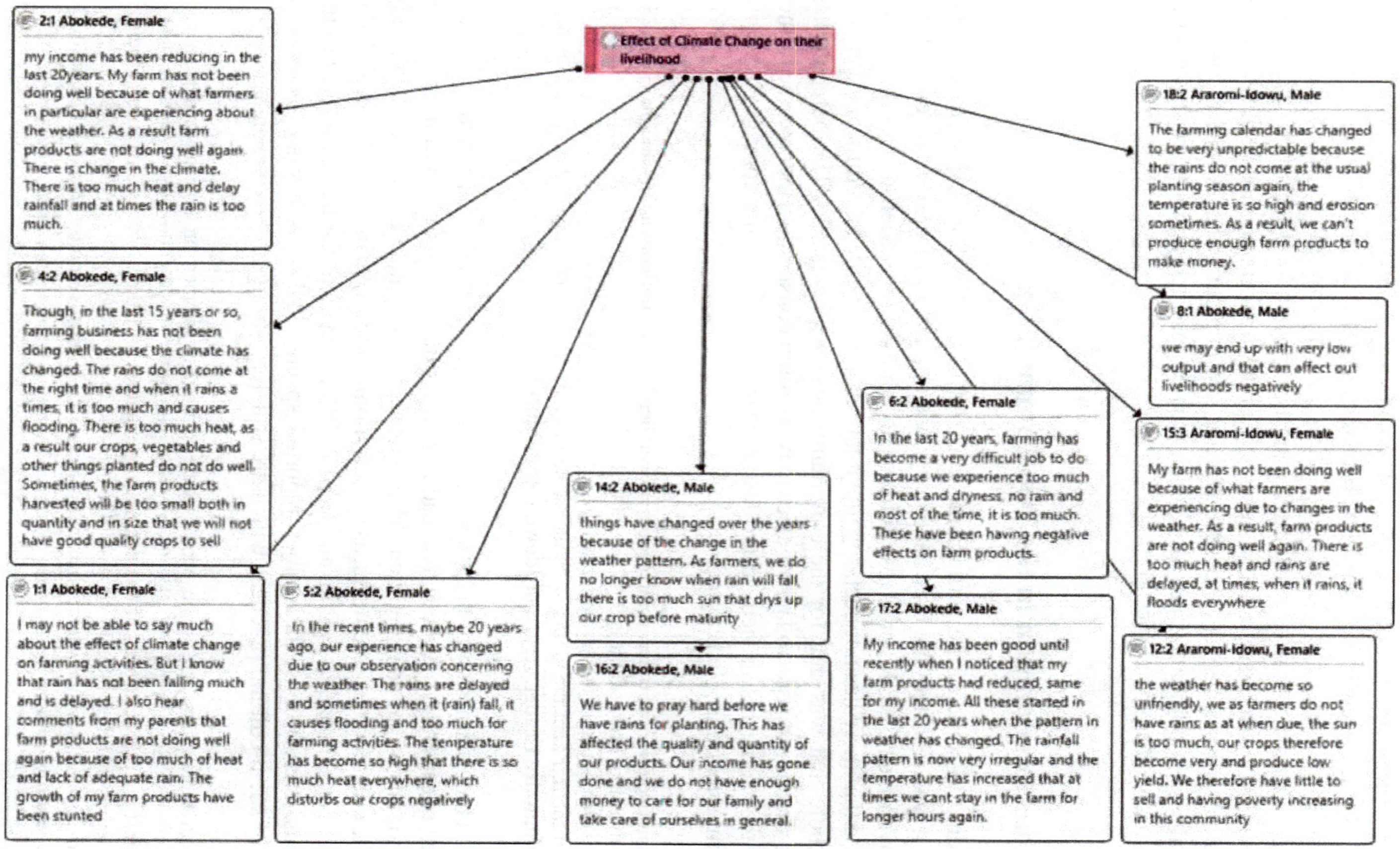

Fig. 4 Effects of climate change on participants' livelihood

result in increasing series of devastating natural disasters and low economic activities and poverty level in the rural areas. Consequently, more people will continue to abandon rural areas that are strongly affected by climate change and migrate to cities. Furthermore, extreme weather events such as floods and heat waves can also pose significant health threats such as diarrheal disease, dehydration, or cardiac complications, and mean temperature changes can alter the range and transmission potential of diseases such as malaria (Akinbami and Momodu 2013, GIZ and MoEF 2009).

Coping Strategies from Communities

During the interview conducted, participants revealed that they have employed some strategies such as diversifying into other business line, use of fertilizer, and wetting of plants from nearby river. It was observed that all of the women and just a few of the men were involved in business diversification. This finding agrees with WEDO (2007) that stated that women have developed specific adaptive strategies to cope with climate change-induced problems in livelihood, health, shelter, energy sectors, etc. in some countries. This suggests that women, if exposed to the right training, can employ their initiatives to develop strategies that will help handle immediate and long-term strategies that will gradually reduce the impacts of climate change in the rural areas. Figure 5 presents various coping strategies the women and some men in the communities visited adopted to alleviate the sufferings associated with effects of climate change. The figure depicts the readiness and willingness of the rural women.

However, the women reported that the socio-cultural climate of their communities hinders their ability to cope. Firstly, the practice around land tenure which prevents women from owning land hinders their ability to diversify. This was well experienced in Araromi-Idowu where the community missed out of oil palm intervention due to lack of land and strong existence of patriarchal system in place.

Climatepreneurship Intervention

The climatepreneurship intervention provided a template which would improve the coping strategies of the participants. Climatepreneurship intervention also exposed participants to the use of natural resources in their community that can be turned to various business activities, such as clay work. This became necessary, in order to create avenues to increase their income generation activities. Also, they were trained on entrepreneurship businesses in the oil palm tree products, such as basket weaving. This was with the aim of introducing them to business diversification concepts.

In order to strengthen participants' coping strategies through proper monitoring, the researcher held series of meetings with the government parastatals in charge of women, community development, and extension officers. This was to enable them get proper understanding of climatepreneurship concept and see to the success and continuity of the intervention through its replication in other communities.

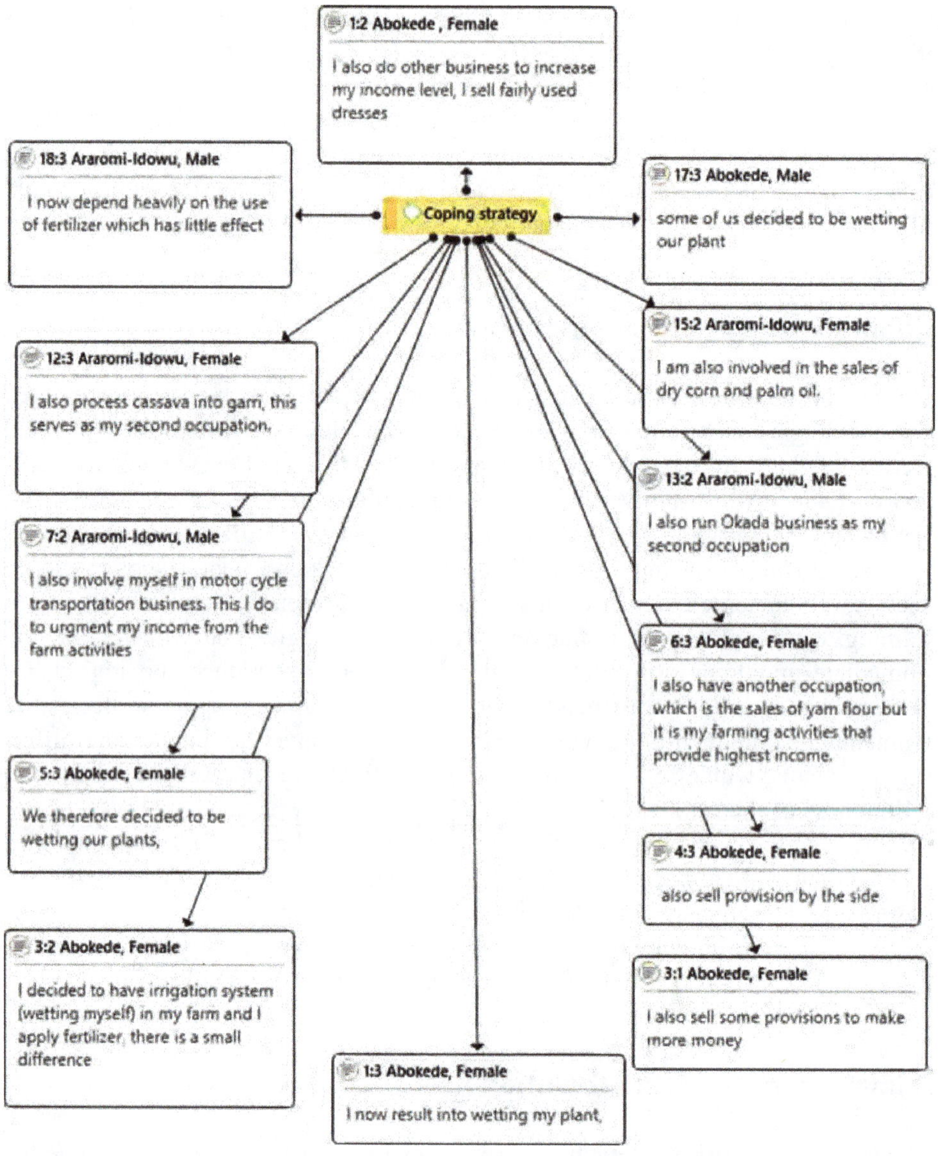

Fig. 5 Coping strategies employed by participants in both Araromi-Idowu and Abokede communities

Effectiveness of Climatepreneurship Intervention

Climatepreneurship was designed by the researcher as an adaptation strategy intervention by employing the opportunities that climate change present to enhance entrepreneurship growth and development for the economic empowerment of the rural men and women. It provided some climate smart grains and seedlings to participants of Araromi-Idowu and Abokede communities in order to empower the women in the communities by helping them to have their livelihood practices improved. This intervention employed some methods that led to its success. Some outcomes were observed as a result of the activities employed by climatepreneurship intervention to demonstrate community communication, partnerships, and bottom-up approach, in order to strengthen their economic capacity and increase entrepreneurship development. These outcomes include:

- The actual needs of participants were known such as lack of monitoring from agricultural extension officers.
- Actual impacts of climate change on their livelihood were stated.
- The drought-resistant maize grains, hybrid oil palm seedlings, and vitamin A-enhanced cassava stems were identified to improve their livelihood practices, thereby enhancing entrepreneurship development and economic empowerment.

The basic and fundamental step was to enlist their interest that led to the acceptance of intervention by community leaders and members. Training sessions were organized in the two communities to educate participants on the handling and management of grains and seedlings. Thereafter, drought-tolerant maize grain varieties, beta-carotene fortified cassava stems, and hybrid oil palm seedlings were distributed to participants and were planted as revealed in Fig. 6. Monitoring tool was employed via periodic visitation to communities and the involvement of the government parastatal attached to agriculture services in the state.

Participants retorted that they have never planted such improved and enhanced seedling varieties before climatepreneurship intervention, though, most of the participants claimed that they have heard of their existence through friends and neighboring communities.

Post-intervention interview was conducted when the drought-tolerant maize grain varieties and beta-carotene cassava stems planted were matured for harvesting, and the participants were able to express their experiences. All the farmers attested to the effectiveness of this intervention over the local maize grains they were used to. Some of the participants from the two communities reported their experiences with climatepreneurship as evidence to its effectiveness and said:

> We have not seen this before. It seems we are going back to those days when farming was rewarding. The maize harvested is quite different from the local grains we normally planted. This one got matured before it was normal 3 months, which means we farmers will have more to sell and thereby have more money to cater for ourselves. It is also sweeter in taste than the one we were used to. What really gladdens my heart is that we will have more trips to the market from the look of things. I will distribute the grains to other people in this community, so that we can all have access to this new grain. **Male Participant, Araromi-Idowu Community, 48 years old**
>
> I never knew the outcome will be this fantastic. We were mocked by some of our colleagues in this community because there was not enough rains. They were so sure that we were wasting our time, despite the training we had before the grains were handed over to us. Maybe, if the researcher had handed the grains over to us without staying on the farm with us to plant every grain, I would have just gone to keep the grains also. But now, I thank God that I participated. Our people that mocked us then are now asking for the grains. I am preparing for good sales this season. **Female Participant, Abokede Community, 50 years old**

Female and male farmers planting drought tolerant maize grains and beta-carotene fortified cassava seedlings

Group of male and female farmers with hybrid oil palm Female farmer planting hybrid oil palm

Distribution of Beta-carotene fortified cassava Both farmers planting drought tolerant maize

Fig. 6 Participants with climate-smart grains and seedlings

> The cassava stem planted yielded so well. Apart from the fact that the tubers are big, I now have cassava that can also improve our health. In this community, we eat a lot of cassava and we process it to fufu (staple food made from cassava) which we take to the city to sell. So apart from having more quantities to sell for money, we are also contributing to the well being of people. I know that we will always have this in this community now by replanting the stem. **Female Participant, Araromi-Idowu Community, 62 years old**

From Fig. 7, it is evident that the rural areas are ready for innovative adaptation strategies, if the right investment is made.

To further enhance the effectiveness of this intervention and strengthen the capacity of participants, it was observed during the pre-intervention activities that

Fig. 7 Post-intervention measure pictures showing harvested maize and cassava in the two communities

Fig. 8 A dug well for improved economic activities by the intervention

the communities lack good clean water supply for the processing of the harvested cassava into fufu (a Nigerian staple food), a well was also dug in the communities to improve their economic activities and their health status because this well water will be used for drinking purposes as well by the communities (see Fig. 8). Unlike the previous intervention of a borehole that was dug by the government which ceased functioning due to the lack of electricity to pump water constantly, the well was made people-friendly which even a young boy or girl would constantly be able to operate manually at any time.

However, in spite of the success of the various aspects of the climatepreneurship intervention measures, the hybrid oil palm aspect was not as successful as others. This is mainly due to the existing land tenure and strong patriarchal system in place in the communities. Since the condition for eligibility for the hybrid oil palm was access to at least 1 acre of land so that the quantity to be harvested would have a positive impact on economic status of the women, most of the women jointly contributed land to be eligible. Also, unfortunately, the community experienced clash between herdsmen and farmers. It was later observed in between the periodical monitoring of the project that cattle rearers had led cattle to the oil palm farm, which was not close to farmers' dwelling places and devastated some of the seedling planted, thereby defeating the objective of this part of the project.

Conclusion

Adapting to climate change represents a new challenge particularly in the rural areas. Already, linkages exist between business-as-usual development strategies and adaptation to climate change impacts. Therefore, development strategies that will generate benefits for managing climate change risks are highly essential. Such will result in poverty reduction, improved nutrition, enhanced education, expanded infrastructure, and improved health which will also reduce vulnerability to climate change and enhance wellness.

This study presented available adaptation strategies suggested by various authors for testing of conformity with rural farmers, following the view that environmental knowledge and resilience to climate change lay within societies and cultures. Farmers in these communities of study have experienced hard times as a result of climate change effects ranging from delayed rainfall, heavy rainfall which lead to flooding, increase in temperature, and drought. Consequently, the communities have suffered some economic hardship and social consequences from impacts of climate change. This necessitates the rural dwellers diversification and integration of other livelihood practices in addition to crop farming. Observations made in the communities of study agreed with previous studies such as Suleiman (2014) and Yusuf et al. (2017).

Due to the low level of education of the rural communities, participants could not process the information received into finding out the causes of climate change and what steps to take to abate the climate change impacts and improve their livelihood practices. It is therefore essential that intensive and extensive sensitization and comprehensive education programs mounted by both public and private sectors including the faith and community-based organizations as well as climate-based NGOs be incorporated into the development intervention strategies in the rural communities in their various local languages for effective and benefitting communication. For example, such sensitization and education programs may include the production of facts sheet and radio interactive programs specifically tailored to address rural farmers' challenges.

Innovative adaptation strategies such as climatepreneurship which consists of activities that are climate smart and make use of locally available resources for economic empowerment and enhancement of the social status of rural women and also help to turn climate change challenges into opportunities have been assessed as effective intervention measures in rural communities. It therefore reveals that successful climatepreneurship intervention measure entails a bottom-up approach that requires a situational report of rural communities for a good understanding of the peculiarities of their climate and environmental assessment based on vegetation zones in each geopolitical zone in the country, as well as locally available raw materials that can be turned to sustainable economic empowerment ventures. These will help to forestall the increasing rural-urban migration with its attendant food insecurity and social vices and further help in developing the rural dwellers especially the womenfolk through sustainable economic empowerment and also enhancing their socio-political status in the society. Ultimately, many more micro-, small-, and medium-sized enterprises (MSME) can begin springing up and thereby enhancing environmentally sustainable industrial rural development and increasing economic value additions to both sub-national and national economic growth and development.

The paper concludes that the purpose of undertaking adaptation strategy is to effectively manage potential climate risks over the coming decades as the effects of climate change continue to increase because such can help inform decisions by farmers (both women and men), agribusiness owners, policy makers, and development partners of implications over a range of timeframes from short-term tactical to long-term strategic options. This is because inadequate consideration of and in-appropriate actions to adaptation options could result in increased vulnerability to climate change generally in the rural areas and particularly by women and children, and thereby giving rise to more severe mitigation targets and costs.

Acknowledgments This research was supported by funding from the Department for International Development (DfID) under the Climate Impact Research Capacity and Leadership Enhancement (CIRCLE) program.

References

Abid M, Scheffran J, Schneider UA, Ashfaq M (2015) Farmers perceptions of and adaptation strategies to climate change and their determinants: the case of Punjab Province, Pakistan. Earth Syst Dyn 6(1):225–243

Adebiyi OA, Adeola AT, Osinowo OA, Brown D, Ng'Ambi JW (2018) Effects of feeding hydroponics maize fodder on performance and nutrient digestibility of weaned pigs. Appl Ecol Environ Res 16(3):2415–2422

Adesina FA, Odekunle TO (2011) Climate change and adaptation in Nigeria: some background to Nigeria's response – III. International conference on environmental and agriculture engineering IPCBEE, vol 15. IACSIT Press, Singapore

Akinbami CAO (2015) Assessment of social dimensions of climate change on Nigerian rural women livelihood practices: implication for entrepreneurship development. Unpublished post-doctoral fellowship report for African Academy of Sciences, Nairobi

Akinbami CAO, Momodu AS (2013) Health and environmental implications of rural female entrepreneurship practices in Osun state Nigeria. AMBIO J Hum Environ. https://doi.org/10.1007/s13280-012-0355-5

Akinbami CAO, Olawoye JO, Adesina FA (2015) Chapter 4: Rural women belief system and attitude toward climate change mitigation and adaptation strategies in Nigeria. In: Walter LO et al (eds) Climate change adaptation, resilience and hazards. Springer Nature, Springer International Publishing AG Switzerland, pp 49–69. https://doi.org/10.1007/978-3-319-39880-8

Akinbami CAO, Olawoye JO, Adesina FA, Nelson V (2019) Exploring potential climate related entrepreneurship opportunities and challenges for rural Nigerian women. J Glob Entrep Res 9:19. https://doi.org/10.1186/s40497-018-0141-3

Ali YO (2011) Legal profession and climate change in Nigeria. http://www.yusufali.net/articles/LEGAL_PROFESSION_CLIMATE_CHANGE_IN_NIGERIA. Accessed 23 Sept 2019

Ariani M, Hervani A, Setyanto P (2018) Climate smart agriculture to increase productivity and reduce greenhouse gas emission– a preliminary study. IOP Conf Ser Earth Environ Sci 200(8):12–24

Ashe MO (2019) International agencies and the quest for food security in Nigeria, 1970–2015. Ubuntu J Confl Transform 8(Special Issue 1):251–274

Berkes F, Folke C (eds) (1998) Linking social and ecological systems: management practices and social mechanisms for building resilience. Cambridge University Press, New York

Bronfenbrenner U (1989) Ecological systems theory. In: Vasta R (ed) Annals of child development, vol 6. Jessica Kingsley, London, pp 187–249

Centers for Disease Control and Prevention (CDC) (2014) The social ecological model: a framework for prevention. http://www.cdc.gov/violenceprevention/overview/social-ecologicalmodel.html. Retrieved 21 Apr 2014

Cutter SL, Barnes L, Berry M, Burton C (2008) A place-based model for understanding community resilience to natural disasters. Global Environ Chang 18:598–606. https://doi.org/10.1016/j.gloenvcha

Doman'ska A, Z˙ukowska B, Zajkowski R (2018) Green entrepreneurship as a connector among social, environmental and economic pillars of sustainable development. Why some countries are more agile? Probl Ekorozwoju 13:67–76

Dube T, Phiri K (2013) Rural livelihoods under stress: the impact of climate change on livelihoods in South Western Zimbabwe. Am Int J Contemp Res 3(5):11–25

Etim NA, Etim NN (2019) Rural farmers' adaptation decision to climate change in Niger Delta region, Nigeria. In: Handbook of climate change resilience. pp 1035–1049. Springer, Cham. https://doi.org/10.1007/978-3-319-93336-8_100

FAO (2011) The global bioenergy partnership sustainability indicators for bioenergy. GBEP

FAO (2018) Climate-smart agriculture. Food and Agriculture Organization of the United Nations

Federal Ministry of Environment of Nigeria (2010) Climate change. Available at http://environment.gov.ng/issues/climate-change/ (accessed on March 12, 2012)

GIZ, MoEF (2009) Adaptation to climate change with a focus on rural areas and India. A publication of German development cooperation institution Deutsche Gesellschaft für Internationale Zusammenarbeit (GIZ) GmbH and Ministry of Envi-ronment and Forests (MoEF), Government of India

Howden SM, Soussana JF, Tubiello FN, Chhetri N, Dunlop M, Meinke H (2007) Adapting agriculture to climate change. PNAS 104(50):19691–19696. www.pnas.org/cgi/doi/10.1073/pnas.0701890104

Hultman M, Bonnedahl KJ, O'Neill KJ (2016) Unsustainable societies – sustainable businesses? Introduction to special issue of small enterprise research on transitional Ecopreneurs. Small Enterp Res 23:1–9

Ikehi ME, Onu FM, Ifeanyieze FO, Paradang PS (2014) Farming families and climate change issues in Niger Delta region of Nigeria: extent of impact and adaptation strategies. Agric Sci 5:1140–1151. https://doi.org/10.4236/as.2014.512124. Scientific Research. http://www.scirp.org/journal/as

Intergovernmental Panel on Climate Change (IPCC) (2015). http://www.ipcc.ch

IPCC (2014) Climate change 2017: impacts, adaptation, and vulnerability, contribution of working group II to the fourth assessment report of the Intergovernmental Panel on Climate Change. Cambridge, UK

Joo SJ, Nixon D, Stoeberl PA (2011) Benchmarking with data envelopment analysis: a return on asset perspective. BIJ 18(4):529–542. https://doi.org/10.1108/14635771111147623

Kirkwood J, Walton Y (2010) What motivates ecopreneurs to start businesses? Int J Entrep Behav Res 16(3):204–228

Kumar A, Sharma P (2013) Impact of climate variation on agricultural productivity and food security in rural India. Discussion paper, No. 2013-43|August 21, 2013. http://www.economics-ejournal.org/economics/discussionpapers/2013-43

Lin TC, Shaner PJ, Wang LJ, Shih YT, Wang CP, Huang GH, Huang JC (2015) Effects of mountain tea plantations on nutrient cycling at upstream watersheds. Hydrology and Earth System Sciences 19(11):4493–4504

MOVE D.5 (2011) D 5. Generic conceptual framework for vulnerability measurement, Methods for the improvement of Vulnerability Assessment in Europe, EUROPEAN COMMISSION DG ENVIRONMENT

Nwokeoma BN, Amad K, Chinedu AK (2017) Climate variability and consequences for crime, insurgency in North East Nigeria. Mediterr J Soc Sci 8(3):171. https://doi.org/10.5901/mjss.2017. MCSER Publishing, Rome-Italy

Omodanisi EO, Egwakhe AJ, Ajike EO (2020) Smart agri-preneurship and farm yield in Nigeria. Global Sci J 8(3):1722–1740. ISSN 2320-9186

Pai A, Shah S, Bohara R (2020) Smart agriculture (No. 2699). EasyChair

Ponisio LC, Ehrlich PR (2018) Diversification, yield and a new agricultural revolution: Problems and prospects. Sustainability 8(11):1118

Redman C, Grove MJ, Kuby L (2004) Integrating social science into the long term ecological research (LTER) network: social dimensions of ecological change and ecological dimensions of social change. Ecosystems 7(2):161–171

Rehman A, Shaikh S (2014) Smart agriculture. Int J Commun Netw Inf Secur 32(2):263–270

Rekik L, Bergeron F (2017) Green practice motivators and performance in SMEs: a qualitative comparative anaysis. J Small Bus Strateg 27:1–18

Sadashivam T (2010) Climate change and rural development: the challenges for agricultural dynamics. Int J Rural Stud (IJRS) 17(2). ISSN 1023–2001. www.vri-online.org.uk/ijrs

Salami A (2010) Climate change mitigation and adaptation options: the Nigeria experience. Available: http://www.spaeloauife.net/develoing.html

Slater R, Peskett L, Ludi E, Brown D (2007) Climate change, agricultural policy and poverty reduction – How much do we know? Natural Resources Perspectives, September, 2007. UK: Overseas Development Institute. ISSN 1356–9228

Suleiman IL (2014) Appraisal of climate change and agriculture in Nigeria. J Geogr Reg Plan 7(9):176–184. https://doi.org/10.5897/JGRP2013.0405

UNDP (2010) Designing climate change adaptation initiatives: a UNDP toolkit for practitioners. UNDP Bureau for Development Policy. UNDP, New York

Women's Environment & Development Organization (2007) Changing the climate: why women's perspectives matter. WEDO, New York

Yilmaz DG (2014) Adaptation of rural communities and understanding their socioeconomic vulnerability for future. 4th international conference on building resilience, building resilience 2014, 8–10 September 2014, Salford Quays, United Kingdom. Procedia Econ Finance 18:536–543

Yusuf OJ, Ayanda IF, Olooto FM, Salawu OL (2017) Livelihood diversification amongst pastoralists and conflict with Arable crop farmers: empirical evidence from Kwara state, Nigeria. Nigerian Journal of Rural Sociology 17(1):45–53

4

Climate Change Adaptation Among Smallholder Farmers in Rural Ghana

Peter Asare-Nuamah and Athanasius Fonteh Amungwa

Contents

Abstract

Climate change has the potential to disrupt sustainable development initiatives, particularly in developing economies. A substantial body of literature reveals that developing economies are vulnerable to climate change, due to high dependency on climate-sensitive sectors, such as agriculture. In Ghana, a growing body of literature has revealed multiple adaptation strategies adopted by smallholder farmers to respond to and reduce climate change impacts. However, there is a dearth of literature on the effectiveness of adaptation strategies. This chapter explores the adaptation strategies of smallholder farmers and analyzed the predictors of effective adaptation. Through the technique of simple random

P. Asare-Nuamah (✉)
Institute of Governance, Humanities and Social Science, Pan African University, Soa, Cameroon

School of Sustainable Development, University of Environment and Sustainable Development, Somanya, Eastern Region, Ghana

A. F. Amungwa
Department of Sociology and Anthropology, Faculty of Social and Management Sciences, University of Buea, Buea, Cameroon

sampling, 378 smallholder farmers were selected, and data was collected using a questionnaire survey. Descriptive and inferential statistics were performed using the SPSS software. The findings indicate that smallholder farmers adopt multiple adaptation strategies to reduce the impact of climate change. In addition, it is revealed that marital status, years of farming experience, knowledge of climate change, and education are significant predictors of adaptation. Moreover, the chapter found that marital status, weedicide application, change in staple food consumption, and planting of early-maturing crops are good predictors of effective adaptation. The chapter recommends the need to intensify adaptation strategies through agricultural extension programs and interventions that improve rural food security and livelihood. In addition, the chapter recommends strengthening the capacity of farmer organizations and rural institutions, particularly agricultural extension and advisory services.

Keywords

Climate change · Climate change adaptation · Effective adaptation · Smallholder farmers · Ghana

Introduction

Background

Climate change adaptation is defined as the adjustment in a system to respond to, recover from, and exploit opportunities, to reduce the impact of climate change. Adaptation has gained center stage in global development debates and discourse (IPCC 2018; World Bank 2018). The heightened attention of adaptation stems from the fact that climate change affects every component of the earth's system, particularly the biosphere and essential life-dependent resources (IPCC 2018). More importantly, climate change affects agriculture, thereby increasing the vulnerability of developing countries, whose economies are largely dependent on climate-sensitive sectors, including agriculture (AGRA 2014; IPCC 2018). According to the Intergovernmental Panel on Climate Change (IPCC) (2014, 2018), adaptation strategies are very important in developing economies in Africa and Asia, as the impact of climate change is already felt in extreme in these regions. For instance, climate-related disasters, such as floods and droughts, are predominant and extreme in Africa and Asia. Moreover, El Nino Southern Oscillation (ENSO) has been reported in Southern and West Africa, with prolonged dry season in Ghana (see Hirons et al. 2018; IPCC 2014, 2018). Consequently, a large body of literature has focused on examining the adaptation strategies adopted particularly by smallholder farmers to respond to these changes, with little or no attention on the effectiveness of adaptation as a conduit for food security and livelihood, which this chapter seeks to address. Dankelman (2010) asserts that adaptation strategies, particularly in developing countries, are ineffective and unsustainable, as they are practiced in a resource-constrained environment, which calls for the need to examine effective adaptation strategies employed by smallholder farmers to enhance their food security and livelihood.

According to the European Environment Agency (EEA 2013), effective adaptation requires holistic, concrete, and process-based measures and approaches, whereas the German Development Agency (GIZ 2014) and IPCC (2007, 2014) argue that effective adaptation strategies have the potential to increase adaptive capacity and offset the negative impact of climate change, by decreasing sensitivity and exposure to climate change. Globally, farmers have adopted adaptation strategies to reduce vulnerability and respond to climate shocks and stresses, such as erratic rainfall, high temperature, flood, drought, and pests, among others (IPCC 2007, 2014, 2018). Adaptation strategies may be self-induced or externally planned; proactive, reactive, or concurrent; planned or spontaneous; and tactical or strategic. In addition, they may occur at different scales, such as local, regional, or international; by agents of adaptation including but not limited to individuals, communities, governments, the private sector, and non-governmental/civil society organizations (Antwi-Agyei et al. 2014a; Bryant et al. 2000; Smit et al. 2000; Smit and Skinner 2002; IPCC 2014). To ensure an effective adaptation, it is essential for adaptation agents to adopt planned adaptation behaviors, which are typical of disaster preparedness, as well as ensure effective response and recovery to disasters (World Bank 2018b).

A growing body of literature has revealed that farmers have adapted to climate change through multiple on-/off-farm adaptation strategies such as climate-smart agricultural practices, to enhance their livelihood and food security, which depend largely on the environment (Antwi-Agyei et al. 2014a; Gyampoh et al. 2009; Pearce and Ford 2015; Yaro et al. 2014). In Ghana, for instance, crop and livelihood diversification; pesticide, weedicide, and fertilizer application; water harvesting; and planting of early-maturing and drought resistant crops have been reported in the literature (Antwi-Agyei et al. 2014a, b; Fosu-Mensah et al. 2012; Gyampoh et al. 2009; Osumanu et al. 2017; Yaro 2013a, b; Yaro et al. 2014). Nevertheless, Sarpong and Anyidoho (2012) argue that the adaptation strategies in Ghana are not effective due to poor policy environment, whereas Connolly-Boutin and Smit (2016) and the IPCC (2014) argue that poor natural resource management, low adaptive capacity, and poverty make the adaptation strategies ineffective in Africa.

According to the GIZ (2014) and Turner et al. (2003), entitlement, access, and endowment influence the adaptive capacity to climate change. This corroborates the assertion that the assets owned, used, and/or available to communities and households enable them to adapt to climate change (IPCC 2007, 2014). Similarly, studies have demonstrated that adaptation is dependent on multiple socioeconomic factors, such as age, gender, years of farming experience, assets, level of education, household size, and technology (Adu et al. 2018; Fosu-Mensah et al. 2012; Lopez-Ridaura et al. 2018). The present study aims to achieve the following objectives: explore adaptation strategies of smallholder famers and predict factors that promote adaptation and effective adaptation. The rest of the paper presents the methods, results, and conclusion.

Description of the Study Area and Methodology

Study Area

The study was conducted in the Adansi North District, located in the Ashanti Region of Ghana on longitude 1.50 W, latitude 1.4 N and longitude 1.5 W, latitude 6.30 N. It covers about 853.63 km^2, which is equivalent to 4.7% of the surface area of the region. It also has a total population of about 107,109, with 54,036 (50.5%) females and 53,055 (49.5%) males (GSS 2010, 2014). The dependency ratio in the district is high, with males found to be more dependent than females (GSS 2014). Located in the semi-equatorial climate, the district experiences high temperature and rainfall, with a mean rainfall of 1250 mm to 1750 mm and temperature of 27 °C. The district has bi-modal rainfall patterns, comprising of major and minor rainy seasons. February is the hottest month in the district, and agriculture is the major economic activity, employing about 77% of the labor force, followed by services (15%) and manufacturing (8%) (GSS 2014). The district has rich ochrosols, which favor agriculture and forest vegetation. The major food crops produced here include cassava, plantain, cocoyam, maize, yam, and cash crops, such as cocoa and oil palm. As a rural district, poverty, low income, and lack of access to basic socioeconomic assets are dominant in the district (GSS 2014).

Research Methods

Study Site Selection and Sampling Method

A total of 15 communities from 7 operational areas were selected via multistage sampling techniques comprising of cluster and simple random sampling. The district has 15 operational areas which served as the clusters from which 7 were randomly selected. A sample of 378 farm household heads was randomly selected using the district census data as the sample frame. Krejcie and Morgan's (1970) statistical table for sample size selection led to the selection of 378 respondents from the household population of 23,863.

Data Sources and Collection Methods

The chapter used quantitative primary data. Data collection lasted for 6 months, starting from April to September 2018. Prior to data collection, the Institutional Review Committee of the Pan African University, Cameroon, approved the study. The data collection instrument was questionnaire survey. The questionnaire was divided into sections to collect data on the respondents' demographic characteristics, access to essential assets, and adaptation strategies.

Data Analysis

Primary data was analyzed using the Statistical Package for Social Sciences (SPSS) version 20. Data cleaning was performed through frequency analysis to identify the missing data. Preliminary analyses, such as outliers and normality tests, were

Table 1 Variables description and coding

Variable description	Variable coding
Dependent variable	
Have you adapted to climate change?	Yes = 1 No = 0
Independent variables	
Sex	Male = 1 Female = 0
Age	Years (Number)
Household size	Number
Education	Yes = 1 No = 0
Marital status	Married = 1 Else = 0
Access to land	Yes = 1 No = 0
Access to extension	Yes = 1 No = 0
Access to market	Yes = 1 No = 0
Knowledge of climate change	Yes = 1 No = 0
Access to technology (e.g. phone)	Yes = 1 No = 0
Farming experience	Yes = 1 No = 0

Source: Fieldwork, 2018

conducted using box and scatter plots as well as line graphs (Pallant 2016). Two binary logistic regression models were specified for the analysis. The first binary logistic regression model explored factors that predicted the respondents' climate change adaptation strategies. The dependent variable in the first model was adaptation that was operationalized as whether or not a respondent has adapted to climate change using any of the strategies available to them. The independent variables comprised of categorical and continuous variables, such as age, sex, household size, farming experience, marital status, access to land, access to technology (phone), access to market, extension service, and knowledge of climate change. Table 1 presents the variable coding and description.

The use of the preliminary regression model without the predictor variables resulted in an overall percentage of 59.2% in terms of whether respondents have adapted or not, which served as a baseline for comparison with the regression model that included predictor variables (Pallant 2016). The Omnibus Tests of Model Coefficient, which determines the "goodness of fit" of the regression model resulted in a significance level of 0.006 (p <0.05). The Omnibus goodness of fit test also resulted in a chi-squared value of 26.193, with 11 degrees of freedom. In addition, the Hosmer–Lemeshow goodness-of-fit test resulted in a chi-squared value of 6.701

with 8 degrees of freedom and a significant value of 0.569. Pallant (2016) asserts that for the Hosmer–Lemeshow test to be significant, the significant value must exceed 0.05. The significant value of 0.569 exceeded 0.05, which is indicative of the goodness of fit of the model. The Cox and Snell R-squared and Nagelkerke R-squared indicated that the predictor variables account for 26.7% and 29.1% of variation in the dependent variable, respectively. In addition, the percentage accuracy in classification (PAC) table (Pallant 2016) resulted in an overall percentage of 62.3%, which exceeded the initial value of 59.2%. This chapter computed for the sensitivity of the model. According to Pallant (2016), the sensitivity of the model is defined as the percentage of the respondents who have or have not adapted to climate change, which has been accurately identified by the model. It is computed by dividing the number of cases predicted by the total number predicted and multiplied by 100 (Pallant 2016). The positive predictive (true positive) value indicates the percentage of respondents who have adapted to climate change and has been accurately identified by the model. The positive predictive value is 57.5%. Conversely, the negative predictive value (true negative) is the percentage of respondents who have not adapted and has been accurately identified by the model. The computation resulted in 61.8% for the negative predictive value.

In the second model, a binary logistic regression analysis was conducted to understand the adaptation strategies that effectively improve the respondents' food security and livelihood. The dependent variable was "whether or not adaptation strategies effectively enhance food security and livelihood of respondents." The independent variables consisted of categorical variables of 24 different adaptation strategies adopted by the respondents. In addition, the predictors of adaptation from the first model were included in the second model. Table 2 presents the coding and description of variables used in the second regression model.

The initial regression model without the predictor variables resulted in an overall percentage of 88.9% in terms of whether or not adaptation enhances food security and livelihood. The Omnibus Test of Model Coefficient resulted in a significance level of 0.018 ($p < 0.05$), with a chi-squared value of 45.858 and 28 degrees of freedom. The Hosmer–Lemeshow goodness-of-fit test resulted in a chi-squared value of 7.085 with 8 degrees of freedom, at a significant value of 0.527. The Cox & Snell R-squared and Nagelkerke R-squared values were 0.114 and 0.227 respectively, indicating that the predictor variables account for 11.4% and 22.7% of the variation in the dependent variable, respectively.

The PAC table resulted in an overall percentage of 89.4%, exceeding the initial value of 88.9%. In addition, the chapter computed for the sensitivity of the model, which is the percentage of the respondents accurately identified by the model as those who indicated that adaptation enhances or does not enhance food security and livelihood. Sensitivity is computed by dividing the number of cases predicted by the total number predicted and multiplied by 100 (Pallant 2016). The positive predictive value (true positive) indicates the percentage of respondents who indicated that adaptation effectively enhances food security and livelihood. The positive predictive value was 89.8%. The negative predictive value (true negatives) is, however, the

Table 2 Variables description and coding

Variable description	Variable coding
Dependent variable	
Does adaptation effectively enhance food security and livelihood?	Yes = 1 No = 0
Independent variables	
Migration	Yes = 1 No = 0
Fertilizer application	Yes = 1 No = 0
So Klin application	Yes = 1 No = 0
Education	Yes = 1 No = 0
Marital status	Married = 1 Else = 0
Delayed farming	Yes = 1 No = 0
Pesticides application	Yes = 1 No = 0
Cocoa pollination	Yes = 1 No = 0
Reduced number of diets	Yes = 1 No = 0
Reduced diet size	Yes = 1 No = 0
Change staple diet	Yes = 1 No = 0
Irrigation (hand/machine)	Yes = 1 No = 0
Extension services	Yes = 1 No = 0
Early maturing crops	Yes = 1 No = 0
Improved crops varieties	Yes = 1 No = 0
Livelihood diversification	Yes = 1 No = 0
Knowledge of climate change	Yes = 1 No = 0
Help from family and friends	Yes = 1 No = 0
Farming experience (years)	Yes = 1 No = 0
Crop diversification	Yes = 1 No = 0
Multiple cropping season	Yes = 1 No = 0
Weedicides application	Yes = 1 No = 0
Cocoa spraying	Yes = 1 No = 0
Farm expansion	Yes = 1

(continued)

Table 2 (continued)

Variable description	Variable coding
	No = 0
Farm Based Organizations	Yes = 1
	No = 0
Chicken droppings application	Yes = 1
	No = 0
Drought resistant crops	Yes = 1
	No = 0
Change in livestock	Yes = 1
	No = 0

Source: Fieldwork, 2018

percentage of respondents who were accurately identified by the model to have indicated that adaptation does not enhance their food security and livelihood. The negative predictive value was 66.7%.

Results and Discussion

Adaptation Strategies Employed by Smallholder Farmers

The respondents in the study have adopted multiple adaptation strategies as presented in Table 3. The dominant adaptation strategy of the respondents is crop diversification, as reported by 93.4% of respondents. Antwi-Agyei et al. (2014a) similarly reported that majority of smallholder farmers in the Bongo District in the Upper East Region of Ghana practice crop diversification, as a response to climate change. Crop diversification is a traditional farming practice where multiple crops are planted on the same plot of land. The system enables farmers to harvest different crops throughout the year, thereby increasing food security. The practice may be influenced by the different maturity periods of crops, nutrient requirement, and demand on the market. Majority of the respondents have resorted to delay farming (70.1%), fertilizer application (68%), planting of early-maturing crops (42.1%), improved crop varieties (33.9%), and multiple-cropping season (51.3%). About 51.1% of the respondents reported of extension services. Studies have reported that extension services enable farmers to adapt to climate change through the transfer of knowledge and emerging farm practices (Boyaci and Yildiz 2016; Morrison and Sarris 2007).

Aside from extension services, the Government of Ghana's agricultural interventions, such as cocoa spraying and cocoa pollination, were also reported by 42.1% and 34.7% of the respondents, respectively, as adaptation strategies. It is interesting to note that some farmers are turning to the application of chicken dropping (17.5%) as an alternative to chemical fertilizer, which they hinted to be expensive. Unique to the findings of this chapter is the application of "So-Klin," a washing detergent used

Table 3 Respondents' adaptation strategies

Indicators	(%)	N (378)
Migration	8.3	33
Fertilizer application	68.0	257
So Klin application	44.4	168
Planting of early-maturing crops	42.1	159
Improved crop varieties	33.9	128
Livelihood diversification	57.4	217
Help from family and friends	40.5	153
Crop diversification	93.4	353
Delayed farming	70.1	265
Multiple-cropping seasons	51.3	194
Pesticide application	34.7	131
Weedicide application	43.2	164
Cocoa pollination	34.7	131
Cocoa spraying	42.1	159
Reduced number of diets	57.4	217
Reduced meal size	66.4	251
Change in diets/staple food	52.4	198
Irrigation (hand/machine)	26.7	101
Extension services	51.1	193
Change in livestock	35.4	134
Planting of drought-resistant crops	18.3	69
Farm expansion	38.4	145
Joining in farm-based organizations	3.7	14
Chicken dropping application	17.5	66

Source: Fieldwork 2018

to control fall armyworm invasion. Some respondents indicated that they use both "So-Klin" and pesticides to control fall armyworms. It is reported that fall armyworms cause significant crop loss, particularly maize, sorghum, and millet (FAO 2018; Wild 2017). In Ghana, about 1.4 million ha of maize and cowpea have been destroyed by fall armyworms, whereas about 6400 ha of cocoa farms have been invaded (African Union 2017; Mpofu 2017; OCHA 2017). Loss of maize from fall armyworms in Africa has been estimated at US$3 billion (African Union 2017). About 57.4% of the respondents have diversified their livelihood. Livelihood diversification has been reported in by Antwi-Agyei et al. (2014a, b) as a conduit to climate change adaptation. Table 3 demonstrates that the respondents have resorted to diet-related adaptation strategies as a response to climate change. For instance, respondents reported reduction in meal size (66.4%), change in diet/staple food (52.4%), and reduction in the number of diets (57.4%), which are consistent with the literature (Giannini et al. 2016; Quaye 2008). Other adaptation strategies include help from family and friends (40.5%), pesticide (34.7%) and weedicide (43.2%)

application, change in livestock (35.3%), farm expansion (38.4%), irrigation (26.7%), and planting of drought-resistant crops (18.3%).

Predictors of Adaptation Strategies

The output of the binary logistic regression analysis is presented in Table 4. The beta (β) values represent the change in the logit of the outcome variable (the log odds ratio) with a unit change in the predictor variable. The β value for years of farming experience, for instance, is −0.616, which indicates that an increase in the experience of a respondent by 1 year will result in a decrease in the adaptation of the respondents to climate change by 0.616. Put differently, the finding indicates that as the years of experience of a respondent increase, there is less likelihood that the respondent will not adapt to climate change. An increase in education increases climate change adaptation by 0.581, whereas access to market and marital status also increase adaptation by 0.178 and 0.583, respectively. Respondents with knowledge of climate change are more likely to adapt to climate change by 1.715. These findings are consistent with the literature (Osumanu et al. 2017; Tessema et al. 2018). For instance, as the level of education increases, farmers are likely to increase their climate change knowledge and become more concerned of the need to adapt. In addition, education increases farmers' knowledge of available and emerging adaptation strategies, including technology and climate-smart agriculture.

The Wald statistic with a chi-squared distribution demonstrates whether or not the β values for predictor variables are significantly different from zero (Mensah 2008). From Table 4, household size, for instance, has a Wald statistic of 2.171, which is significantly different from 0. However, the contribution of household size to the model is statistically insignificant at a p-value of 0.141 (thus, p <0.05). From the model, years of experience, education, marital status, and knowledge of climate

Table 4 Binary logistic regression of the predictors of climate change adaptation

Variables	β	Wald	Significance	Exp(β)
Sex	−0.112	0.176	0.675	0.894
Age	0.008	0.728	0.393	1.008
Household size	0.037	2.171	0.141	1.037
Experience	−0.616	3.752	0.050*	0.540
Education	0.581	4.738	0.030*	1.789
Marital status	0.583	4.085	0.043*	1.792
Access to land	−0.520	1.582	0.208	0.594
Access to technology	−0.443	1.712	0.191	0.642
Access to extension services	−0.028	0.012	0.914	0.973
Market	0.178	0.451	0.502	1.195
Climate change knowledge	1.715	4.739	0.029*	5.559
Constant	−2.310	5.263	0.022	0.099

Source: Computed from fieldwork 2018

change have Wald statistics of 3.752, 4.738, 4.085, and 4.739, respectively, which are statistically significant at p-values of 0.050, 0.030, 0.043, and 0.029, respectively. Marital status, for instance, plays a significant role in climate change adaptation, as it promotes pool of resources. In addition, married household heads are more likely to adapt due to their responsibility to feed their families and cater for their wards. Studies have revealed that the farming experience and knowledge of climate change significantly influence climate change adaptation (Osumanu et al. 2017; Tessema et al. 2018).

Another important statistic presented in Table 4 is the Exp(β). These values, which are the odd ratios of the predictor variables, show the odds of the outcome variable occurring as a result of a unit change in the predictor variable. Statistically, the odd of an event is the probability of that event occurring divided by the inverse or the probability of that event not occurring (Mensah 2008). Exp(β) values greater than 1 means that an increase in the predictor variables results in an increase in the odds of the outcome variable, whereas Exp(β) values less than 1 show an inverse relationship between the odds of the outcome and the predictor variable. For instance, knowledge of climate change has Exp(β) value of 5.559, which indicates that there is a positive relationship between adaptation and knowledge of climate change. As such, it can be interpreted that increasing the knowledge of climate change of a respondent increases the respondent's adaptation to climate change. Table 4 demonstrates that age, household size, education, marital status, and market increase adaptation to climate change, whereas sex, experience, access to land, technology, and extension service reduce adaptation to climate change. It can be concluded from the model that years of experience, education, marital status, and knowledge of climate change are significant predictors of climate change adaptation, which is consistent with the previous studies (Osumanu et al. 2017; Tessema et al. 2018). The equation of the model can be written as Logit(Adaptation) = −2.310(Constant) −0.112*Sex + 0.008*Age + 0.037*Household size −0.616*Experience + 0.581*Education + 0.583*Marital status − 0.520*Land − 0.443*Technology − 0.028*Extension + 0.178*Market + 1.715*Climate change knowledge.

Predictors of Effective Adaptation

Table 5 presents the output of the binary logistic regression estimation. The β value is the change in the logit of the outcome variable (the log odds ratio) as a result of a unit change in a the predictor variable. The β values indicate that improved crop varieties, weedicide application, and change in staple food consumption, for instance, increase food security and livelihood by 1.130, 1.315, and 1.333, respectively. It can be asserted that improved crop varieties, weedicide application, and change in staple food consumption have positive relationships with food security and livelihood. The more a respondent adopts these adaptation strategies, the more likely food security and livelihood of the respondent will improve. Other adaptation strategies, such as cocoa spraying, multiple-cropping season, and drought-resistant crops, are positively related to food security and livelihood as a unit increase in these strategies

Table 5 Binary logistic regression of the predictors of effective adaptation

Variables	β	Wald	Significance	Exp(β)
Migration	0.305	0.121	0.728	1.357
Fertilizer application	0.016	0.001	0.973	1.016
So Klin application	0.418	0.586	0.451	1.519
Planting of early-maturing crops	−1.325	3.918	0.048*	0.266
Improved crop varieties	1.130	2.374	0.123	3.095
Livelihood diversification	−0.705	2.212	0.145	0.494
Help from family and friends	0.221	0.193	0.660	1.248
Crop diversification	0.246	0.077	0.782	1.279
Delayed farming	−0.655	1.664	0.197	0.520
Multiple-cropping season	0.531	0.922	0.337	1.701
Pesticide application	−1.117	2.138	0.144	0.327
Weedicide application	1.315	3.444	0.500*	3.724
Cocoa pollination	0.200	0.068	0.795	1.222
Cocoa spraying	0.474	0.504	0.478	1.607
Reduced number of diets	−1.156	3.432	0.064	0.315
Reduce meal size	−0.357	0.292	0.589	0.700
Change in staple food consumption	1.333	7.111	0.008*	3.793
Irrigation	−0.308	0.238	0.626	0.735
Extension services	0.176	0.160	0.689	1.192
Change in livestock	−0.843	2.735	0.098	0.430
Planting of drought-resistant crops	0.884	1.429	0.232	2.420
Farm expansion	−0.008	0.000	0.987	0.992
Joining in farm-based organization	−1.145	1.506	0.220	0.318
Chicken dropping application	−0.822	3.465	0.063	0.440
Years of farming experience	0.154	0.100	0.752	1.166
Education	−0.365	0.692	0.405	0.694
Marital status	0.749	3.582	0.503*	2.116
Knowledge of climate change	−18.547	0.901	0.998	0.001
Constant	20.933	2.385	0.998	1.681

Source: Computed from fieldwork 2018

improves food security and livelihood by a margin of 0.474, 0.531, and 0.884, respectively. Among the predictors of adaptation, years of farming experience (0.154) and marital status (0.749) have a positive relationship with food security and livelihood.

The Wald statistic demonstrates whether or not the β values for predictor variables are significantly different from zero (Mensah 2008). Predictor variables such as marital status, chicken dropping application, planting of early-maturing crops, livelihood diversification, pesticide application, weedicide application, reduction in the number of diets, and change in staple food consumption, for instance, are significantly different from zero with Wald statistics of 3.582, 3.465, 3.918, 2.212, 2.138, 3.444, 3.432, and 7.111, respectively. The significance column in the table presents the level of significance of the predictor variables in predicting effective adaptation. The model indicates that effective adaptation is significantly predicted by planting early-maturing crops (0.048), weedicide application (0.050), change in staple food consumption (0.008), and marital status (0.053). The Exp(β) values are the odds of the outcome variable occurring as a result of a unit change in the predictor variable. Statistically, the odd of an event is the probability of the event occurring divided by the inverse or the probability of that event not occurring (Mensah 2008). Exp(β) values greater than 1 indicate that an increase in the predictor variables result in an increase in the odds of the outcome variable, whereas Exp(β) values less than 1 show an inverse relationship between the odds of the outcome and the predictor variable.

For instance, planting of drought-resistance crops, change in staple food consumption, weedicide application, and improved crop varieties have Exp(β) values greater than 1 which indicates a positive correlation between effective adaptation (food security and livelihood) and these predictors. Marital status (0.001), planting of early-maturing crops, (0.266), and pesticide application (0.372), for instance, have Exp(β) values less than 1 which indicates an inverse or negative relationship with effective adaptation (food security and livelihood). It can be summed up from the model that planting of early-maturing crops, weedicide application, change in staple crop consumption, and marital status significantly predict effective adaptation (food security and livelihood). The equation of the model can be written as Logit(EffectiveAdaptation) = 20.933(Constant) + 0.035*Migration + 0.016*Fertilizer application + 0.418*So Klin application −1.325*Planting of early-maturing crops + 1.130*Improved crop varieties −0.705*Livelihood diversification + 0.221*Help from family and friends + 0.246*Crop diversification −0.655*Delayed farming + 0.531*Multiple-cropping season −1.117*Pesticide application + 1.315*Weedicide application + 0.2008*Cocoa pollination + 0.474*Cocoa spraying −1.156*Reduced number of diets − 0.357*Reduced meal size + 1.333*Change in staple food − 0.308*Irrigation + 0.176*Extension − 0.843*Change in livestock + 0.884*Planting of drought-resistant crops − 0.008*Farm expansion − 1.145*Joining in farm-based organization − 0.822*Chicken dropping application + 0.154*Experience − 0.365*Education +0.749*Marital status − 18.547*Climate change knowledge.

Conclusion and Recommendations

Ghana's ability to promote food security, livelihood, and rural and pro-poor economic development is largely dependent on agriculture and hence smallholder farmers who constitute the bulk of the labor force in Ghana and contribute about 80% of the food produced. The agriculture sector, in particular, and smallholder farmers, in general, are vulnerable to climate change, due to subsistence and climate-dependent nature of agriculture and low adaptive capacity of smallholder farmers, in addition to high poverty and less access to essential assets. Nevertheless, smallholder farmers in Ghana are adapting to climate change through multiple self-induced and externally supported adaptation strategies. This chapter, which sought to explore smallholder famers' adaptation and examine the predictors of adaptation and effective adaptation, found that smallholder farmers in the rural Adansi North District of Ghana use multiple on-/off-farm and diet-based adaptation strategies to respond to climate change. The chapter also found that marital status, years of farming experience, education, and knowledge of climate change significantly influence adaptation, whereas marital status, weedicide application, change in diet/staple food, and planting of early-maturing crops significantly enhance effective adaptation (food security and livelihood) of smallholder farmers.

While farmers are adapting to climate change, most adaptation strategies of smallholder farmers do not necessarily enhance their food security and livelihood, which may be due to the resource-constrained environment and low adaptive capacity. As such, it is imperative that social intervention programs and policies that promote resource distribution and access must be strengthened, particularly in rural communities, to increase smallholder farmers' access to essential socioeconomic assets, thereby increasing adaptive capacity, improving adaptation strategies, and reducing climate change impacts. In addition, mass education of farmers on climate change must be intensified to improve climate change knowledge/perception, reduce misconceptions associated with climate change, and ensure effective adaptation, as farmers' climate change perception strongly influences climate change adaptation. Farm-based organizations can play a significant role in climate change education and must be promoted particularly in rural communities. Moreover, local institutions, particularly those that provide extension and advisory services, must be intensified and resourced with technology, personnel, and infrastructure to provide services that enable farmers to adapt effectively to climate change through skills and knowledge transfer.

While this chapter limited its geographic scope to only the Adansi North District, the chapter could do better if it covered a wider geographic scope across different ecological zones in Ghana, to deepen the understanding of how different communities are adapting to climate change, which is very important for fit-for-purpose climate change policies. Hence, further studies must consider larger population with a wider geographic location to better understand climate change adaptation in Ghana. In addition, a combination of quantitative and qualitative data on climate change adaptation can greatly improve an understanding of climate change and advance the debate on adaptation. This chapter used only quantitative data, but future studies must go further to include qualitative lived experiences of smallholder farmers. Nevertheless, the chapter has made significant contributions to climate change discourse and has provided insights, with the potential to shape climate change policies in Ghana, in particular, and Africa, in general.

Acknowledgments The authors are grateful to the African Union for funding this study through the Pan African University PhD research grant awarded to the first author.

References

Adu DT, Kuwornu JKM, Anim-Somuah H, Sasaki N (2018) Application of livelihood vulnerability index in assessing smallholder maize farming households' vulnerability to climate change in Brong-Ahafo region of Ghana. Kasetsart J Soc Sci 39(1):22–32. https://doi.org/10.1016/j.kjss.2017.06.009

African Union (2017) Phytosanitary news bulletin: fall armyworms storm Africa. African Union, Yaounde

AGRA (2014) Africa agriculture status report 2014: climate change and smallholder agriculture in sub-Saharan Africa. AGRA, Nairobi. Retrieved from www.agra.org

Antwi-Agyei P, Dougill AJ, Stringer LC (2014a) Barriers to climate change adaptation: evidence from northeast Ghana in the context of a systematic literature review. Clim Dev 7:297–309. https://doi.org/10.1080/17565529.2014.951013

Antwi-Agyei P, Stringer LC, Dougill AJ (2014b) Livelihood adaptations to climate variability: insights from farming households in Ghana. Reg Environ Chang 14(4):1615–1626. https://doi.org/10.1007/s10113-014-0597-9

Boyaci M, Yildiz O (2016) An overview of agricultural extension services in Turkey. Bulgarian J Agr Sci 22(1):151–157

Bryant C, Smit B, Brklacich M, Johnston T, Smithers J, Chiotti Q, Singh B (2000) Adaptation in Canadian agriculture to climatic variability and change. Clim Chang 45:181–201

Connolly-Boutin L, Smit B (2016) Climate change, food security, and livelihoods in sub-Saharan Africa. Reg Environ Chang 16:385–399. https://doi.org/10.1007/s10113-015-0761-x

Dankelman I (2010) Gender and climate change: an introduction. Routledge, London

EEA (2013) Adaptation in Europe: addressing risks and opportunities from climate change in the context of socioeconomic developments. EEA, Copenhagen. Retrieved from http://www.eea.europa.eu/publications/adaptation-in-europe

FAO (2018) Briefing note on FAO actions on fall armyworm in Africa. FAO, Rome. Retrieved from http://www.fao.org/3/BS183E/bs183e.pdf

Fosu-Mensah BY, Vlek PL, MacCarthy D (2012) Farmers' perception and adaptation to climate change: a case study of Sekyedumase district in Ghana. Environ Sustain Dev 8(3):1–13. https://doi.org/10.1007/s10668-012-9339-7

Giannini A, Krishnamurthy PK, Cousin R, Labidi N, Choularton RJ (2016) Climate risk and food security in Mali : a historical perspective on adaptation. Earth's Future 5:144–157. https://doi.org/10.1002/eft2.176

GIZ (2014) The vulnerability sourcebook: concepts and guidelines for standardized vulnerability assessment. GIZ, Eschborn

GSS (2010) Population and housing census 2010. GSS, Accra

GSS (2014) 2010 population and housing census: district analytical report, Adansi North District. GSS, Accra

Gyampoh BA, Amisah S, Idinoba M, Nkem JN (2009) Using traditional knowledge to cope with climate change in rural Ghana. Unasylva 60(231/232):70–74. https://doi.org/10.1016/j.foodchem.2011.01.039

Hirons M, Boyd E, McDermott C, Asare R, Morel A, Mason J, . . . Norris K (2018) Understanding climate resilience in Ghanaian cocoa communities – advancing a biocultural perspective. J Rural Stud 63(August):120–129. https://doi.org/10.1016/j.jrurstud.2018.08.010

IPCC (2007) Climate change 2007: impacts, adaptation and vulnerability. Contribution of working group II to the fourth assessment report of the intergovernmental panel on climate change. Cambridge University Press, Cambridge, MA

IPCC (2014) Climate change 2014: synthesis report. Contributions of working groups I, II and III to the fifth assessment report of the intergovernmental panel on climate change. IPCC, Geneva

IPCC (2018) IPCC special report on the impacts of global warming of 1.5 °C – summary for policy makers. IPCC, Incheon. Retrieved from http://www.ipcc.ch/report/sr15/

Krejcie RV, Morgan DW (1970) Determining sample size for research activities. Educ Psychol Meas 30:607–610. https://doi.org/10.1177/001316447003000308

Lopez-Ridaura S, Frelat R, van Wijk MT, Valbuena D, Krupnik TJ, Jat ML (2018) Climate smart agriculture, farm household typologies and food security: an ex-ante assessment from Eastern India. Agric Syst 159:57–68. https://doi.org/10.1016/j.agsy.2017.09.007

Mensah J (2008) Religious transnationalism among Ghanaian immigrants in Toronto: a binary logistic regression analysis. Can Geogr 52(3):309–330. https://doi.org/10.1111/j.1541-0064.2008.00215.x

Morrison J, Sarris A (2007) WTO rules for agriculture. FAO, Rome

Mpofu V (2017) The foreign invader costing African farmers $3 billion. Retrieved September 15, 2017, from http://www.irinnews.org/news/2017/09/14/foreign-invader-costing-african-farmers-3-billion?utm_source=IRIN+-+the+inside+story+on+emergencies&utm_campaign=860dc2c80d-RSS_EMAIL_CAMPAIGN_ENGLISH_AFRICA&utm_medium=email&utm_term=0_d842d98289-860dc2c80d-7546666

OCHA (2017) West and Central Africa: weekly regional humanitarian snapshot. New York. Retrieved from https://reliefweb.int/sites/reliefweb.int/files/resources/ExternalWeekly24-30October2017.pdf

Osumanu IK, Aniah P, Yelfaanibe A (2017) Determinants of adaptive capacity to climate change among smallholder rural households in the Bongo District, Ghana. Ghana J Dev Stud 14 (2):142–163. https://doi.org/10.4314/gjds.v14i2.8

Pallant J (2016) SPSS survival manual – a step by step guide to data analysis using SPSS program, 6th edn. McGraw-Hill, London

Pearce LBT, Ford JD (2015) Systematic review approaches for climate change adaptation research. Reg Environ Chang 15:755–769. https://doi.org/10.1007/s10113-014-0708-7

Quaye W (2008) Food security situation in northern Ghana, coping strategies and related constraints. J Agric Res 3(5):334–342. https://doi.org/10.1126/science.1185383

Sarpong DB, Anyidoho NA (2012) Climate change and agricultural policy processes in Ghana (Working paper 045). Future Agricultures, Accra

Smit B, Skinner M (2002) Adaptation options in agriculture to climate change: a typology. Mitig Adapt Strateg Glob Chang 7:85–114

Smit B, Burton I, Klein RJ, Wandel J (2000) An anatomy of adaptation to climate change and variability. Climate Change 45:223–251

Tessema YA, Joerin J, Patt A (2018) Factors affecting smallholder farmers' adaptation to climate change through non-technological adjustments. Environ Dev 25:33–42. https://doi.org/10.1016/j.envdev.2017.11.001

Turner RB, Kasperson RE, Matson PA, McCarthy JJ, Corell RW, Christensen L, . . . Schiller A (2003) A framework for vulnerability analysis in sustainability science. Natl Acad Sci 100 (14):8074–8079. https://doi.org/10.11764/j.issn.1672-1926.2014.S1.0091

Wild S (2017) Invasive pest hits Africa. Nature 509:13–14. https://doi.org/10.1038/509015a

World Bank (2018) 2025 targets to step up climate action. World Bank, Washington, DC. Retrieved from http://pubdocs.worldbank.org/en/368601543772742074/2025-Targets-to-Step-Up-Climate-Action.pdf

World Bank (2018b) Adaptation and resilience: action plan. World Bank, Washington, DC. Retrieved from www.worldbank.org/climatechange

Yaro JA (2013a) Building resilience and reducing vulnerability to climate change: implications for food security in Ghana. Friedrich Ebert Stiftung, Accra

Yaro JA (2013b) The perception of and adaptation to climate variability/change in Ghana by small-scale and commercial farmers. Reg Environ Chang 13(6):1259. https://doi.org/10.1007/s10113-013-0443-5

Yaro JA, Teye J, Bawakyillenuo S (2014) Local institutions and adaptive capacity to climate change/variability in the northern savannah of Ghana. Clim Dev 7(3):37–41. https://doi.org/10.1080/17565529.2014.951018

5

Urban Flooding, Adaptation Strategies and Resilience: Case Study of Accra, Ghana

Kwadwo Owusu and Peter Bilson Obour

Contents

Abstract

Despite massive flood controlling investments, perennial flooding continues to be a major challenge in the Greater Accra Metropolitan Assembly in Ghana. Previous studies have mostly considered the vulnerability of Accra to flooding induced by urbanization and climate change. This chapter examined the impacts of and adaptation strategies to flooding in two flood-prone residential areas in Accra. A survey was conducted among 320 household heads to ascertain local impacts of floods and community adaptation strategies. To obtain a broader picture of government interventions and challenges, key stakeholders such as personnel from ministries, departments, and agencies who are involved in city planning, and

K. Owusu · P. B. Obour (✉)
Department of Geography and Resource Development, University of Ghana, Legon, Ghana
e-mail: kowusu@ug.edu.gh; pbobour@ug.edu.gh

private urban planning consultants were interviewed. The study found that a notable driver of floods in Accra is blocked waterways, and flawed and ad hoc engineering works. About three-quarters of the households interviewed have suffered flood-related losses over the past decade such as housing damage, income, and even a death of a relative. Key flood control interventions included dredging prior to start of rains and sporadic demolition of unauthorized buildings on or near waterways to allow free flow of water. However, these interventions only seem to be ephemeral due to the rapid rate of littering and re-siltation of the waterways after few rain events. The study highlights the need for more pragmatic and robust engineering solutions to build resilience of Accra to floods.

Keywords

Barriers to adaptation · Flood control interventions · Perennial flooding · Socio-economic livelihoods

Introduction

Urban flooding is one of the major hydro-climatic disasters which is increasing in frequency and intensity and affecting regions worldwide. Flooding causes serious economic, social, health, cultural, and environmental impacts that affect sustainable development (Kim et al. 2017), particularly in developing regions because of high poverty levels, and poor urban planning and building architecture. Different types of floods can be identified based on the source of water and the process giving rise to the flooding event (Viglione and Rogger 2015). This study focuses on urban floods occurring independently and/or in conjunction with flash and river floods. As at the year 2000, global urban land susceptible to floods was estimated to be around 195,493 km^2 (Güneralp et al. 2015). Africa is one of the continents highly affected by flood-induced disasters, particularly in urban areas. The vulnerability of African cities to urban flooding is partly due to the fast rate of uncontrolled urbanization and its associated problems which constrained flow of water (Amoako 2012; Aboagye 2012; Dewan 2015).

In Ghana, flooding is among the top 10 natural disasters and hazards and ranks second after epidemics such as bacterial and viral infectious diseases, affecting over three million people and causing huge economic losses each year (Okyere et al. 2012). The country has experienced increasing flood events over the years – the severest floods in the last two decades occurred in the years 2007, 2010, 2011, 2015, and 2020, which hit many parts of the country, particularly major urban areas. For instance, on June 22, 2010, 35 bodies were retrieved from flood waters across the country by volunteers and rescue workers (Daily Graphic 5 June, 2015), and on June 3, 2015, flooding with an associated fire killed about 200 people in the commercial hub of Accra (BBC 2015). The consequences of flood-induced disasters in urban areas include loss of properties and lives as outlined above, which also have short- and long-term adverse consequences on urban poverty and development.

Global climate change impacts are likely to increase hydro-climatic disasters like flooding in urban areas. The recent Intergovernmental Panel on Climate Change (IPCC) Special Report on the impacts of global warming of 1.5 °C above preindustrial levels and related global greenhouse gas emission pathways projects increase in the frequency and magnitude of flood due to changing precipitation patterns (IPCC 2018). Despite the large uncertainties in the magnitudes of climate change impacts on water-related shocks, increased frequency of extreme rainfall events will pose a serious threat to urban resilience and sustainability in many developing countries. The high vulnerability of these regions to extreme climatic events results from their low resilience and weak adaptive capacity (IPCC 2007). Adaptation to flood-related impact is not an option but a necessity, which must be embraced across all spheres of planning. However, while the vulnerability of urban areas to flood-related disasters is largely studied, the local impacts and, more importantly, multilevel flood adaptation strategies are poorly understood especially in developing countries.

To contribute to a better understanding of climate change impact on urban areas in Ghana to disasters, and identify key adaptation challenges to guide decision making and for prioritizing city planning needs, a study was conducted in two flood-prone coastal metropolitan areas of Accra and Sekondi-Takoradi. Part of the results recently published by Owusu et al. (2019) investigated the physical and social vulnerability of the two urban cities to climate-induced disasters. Key findings highlight the need to prioritize perennial flooding to safeguard lives and property and to strengthen the cities' resilience to flood-related disasters. Following up on the previous findings by Owusu et al. (2019), the present chapter investigates incidence of flooding, adaptation strategies, and resilience in Accra. Specifically, the study (i) examined impact of floods on socio-economic livelihoods and (ii) explores multi-level flood control measures and adaptation strategies, and associated shortfalls, that is, why interventions to tackle flooding in Accra have been ineffective? It is hoped that a better understanding of flood control measures and adaptation practices and pitfalls would provide essential information on successful adaptation options and help to identify the interventions that need to be improved to build the resilience of Accra and other cities in Ghana to flood-related disasters.

Accra's Vulnerability to Perennial Floods

The term vulnerability has been used by different authors to connote different meanings. Beyond the multidimensional usage of the term vulnerability, most comprehensive and accepted definition in the context of climate change has been given by the IPCC (2000) as: "the degree to which a system is susceptible to, or unable to cope with adverse effects of climate change, including climate variability and extremes." Thus, vulnerability of any system has generally been accepted as a function of three elements: exposure to a hazard, sensitivity to that hazard, and the capacity of the system to cope with and adapt or recover from the effects of those conditions (Jamshidi et al. 2019). Here, vulnerability is used to encompass both the exposure and susceptibility of population to floods. Thus, the chapter examines the exposure of residents in Accra to flooding and the ensuing burden on socio-economic livelihoods and sustainable development in the city.

Accra is a primate city in Ghana with a long history of flood-related disasters as illustrated in Table 1. Records of major flood incidence in the city date back to June 1959 (Daily Graphic 5 June, 2015). More than 16 deadly flood disasters have been recorded since 1959, all of which have caused severe damages to life and property. The vulnerability of Accra to flooding has been attributed to natural factors and processes like its low-lying topography, increase rate of coastal erosion, frequency of torrential rainfall in recent years, combined with anthropogenic factors such as rapid urbanization, chocked drains, and blocked watercourses (Jamshidi et al. 2019; Asumadu-Sarkodie et al. 2015; Appeaning Addo and Appeaning Addo 2016). For instance, on July 3, 1995, the Ghana Meteorological Agency (GMet) recorded 243 mm of rain with a peak intensity of 64 mm in 12 min. The rain lasted for approximately 5 h resulting in flooding in Accra and its environs (Kwarteng 2007) and on May 18, 2008, and October 25, 2011, ~156 mm of rain water was recorded in less than 10 h resulted in a deadly flood in many parts of Accra (The Weekend Globe 2013). According to Maskey and Trambauer (2015), the main process leading to flooding is precipitation falling over an extended period of time in an area which produces runoff due to saturation (antecedent soil moisture) and excess infiltration. The relationship between heavy downpours and incidence of flooding (due to rapid peak flows) has been explained using run-off models (e.g., Asumadu-Sarkodie et al. 2015). Jahanbazi and Egger (2014) stress that extreme rainfall leads to flooding when urban drainage system fails to drain the total amount of rainwater, resulting in what could be termed as an urban pluvial.

Considering that changes in precipitation patterns and extreme rainfall associated with climate change will increase the vulnerability of cities to flooding in many regions (Willems et al. 2012), flooding in Accra will continue to be a major socio-economic and development challenge more than ever. A better understanding of management of flooding and failures is vital to develop sustainable solutions and building the resilience in flood-prone areas in Accra.

Description of the Study Area and Methodology

Accra is the largest metropolitan area in Ghana and has been the administrative capital city of the country since 1877. It has a total land area of 201 km^2 (UN-Habitat 2009). The city has experienced a rapid growth and it is one of the fastest-growing cities in West Africa with an estimated annual average growth rate (2010–2020) of 2.2% (www.macrotrends.net 2020). Accra has physically developed very quickly particularly along the arteries of the major road network radiating from the city center to the countryside (Møller-Jensen et al. 2005). Another worrying trend is the increasing physical development along the low-lying coastal and flood-prone areas of Accra threatening the ecology of coastal lagoons and estuaries (Owusu et al. 2019). The population in the city has increased from about 177,000 inhabitants in 1950s to about 2.5 million inhabitants by the close of 2019 (www.macrotrends.net

Table 1 Magnitude of daily precipitation and flood events and the number of affected people in Accra (1959–2020)

Date	Precipitation-a day before floods (mm)	Precipitation-flood day (mm)	Monthly total (mm)	% of precipitation on flood day[a]	Number of people affected[b]	Reference
June 1959[c]	–	192	–		–	
June 27, 1960	0	98	371	26	–	(Daily Graphic 2015)
September 29, 1963	0	97	190	51	–	
July 4, 1968[d]	5	33	372	9	>1000	(Asumadu-Sarkodie et al. 2015)
June 22, 1973	0.3	175	411	43	–	
July 14, 1991	2	157	263	60	–	
July 3, 1995	0	244	274	89	>1000	(Kwarteng 2007)
June 13, 1997	38	114	353	32	1000s	(Asumadu-Sarkodie et al. 2015)
June 27, 2001	27	81	247	33	100,000	(Aboagye 2012; CNN World 2001)
June 9, 2002	0	123	421	29	~1500	(United Nations Office for the Coordination of Humanitarian Affairs (OCHA) 2002)
March 26, 2007	0	59	63	95	–	
May 18, 2008	0	151	395	38	–	
October 25, 2011	58	98	184	53	43,000	(The Weekend Globe 2013)
June 3, 2015[e]	–	–	–	–	>10, 000	(Daily Graphic 5 June, 2015)
June 8, 2020	–	–	–	–	>50, 000	(Darko 2020)

Source of precipitation data: Ghana Meteorological Agency (GMet), Accra.
[a]Precipitation on the day of floods as a percentage of the monthly total precipitation (mm)
[b]People requiring immediate assistance during an emergency situation plus those injured and homeless
[c]Reported amount of rainfall from Daily Graphic (5 June, 2015)
[d]Precipitation of 148.3 mm was recorded 2 days before day of flood
[e]Flood ignited fire flames in a fuel station. Number of lives lost included victims killed by the fire

2020) and accommodates approximately 8.4% of the ~30 million people in Ghana. Unfortunately, infrastructure and service provision is behind the physical development of Accra and has placed pressure on housing and worsened the already high housing deficit in the city (Rain et al. 2011). As a result, there are many areas that are unplanned or where inadequate planning has been carried out by the land-owners who attempt to integrate their development into the statutory land use plan of Accra leading to a proliferation of slums with about 38% of the city's population living in these underserved settlements (UN-Habitat 2011).

The physical location of Accra in the dry coastal foothills of the Akwapim ranges renders the city vulnerable to flooding. Accra is mostly low lying with average elevation around 20 m and bordered by few hills to the west and northwest. It is drained by many rivers like the Odaw and Dzorwulu that originate from the Akwapim hills. Rainfall in the Southeast of Ghana where Accra is located has high variability (with coefficient of variation of above 30%) which results in high frequent of rainfall extremes (Owusu and Teye 2015). Two residential classes in Accra, namely, second and third areas prone to perennial flooding (Owusu et al. 2019), were selected for the study. The specific neighborhoods were: second class residential areas (Kaneshie, Abelemkpe, and Asylum Down) and third class residential areas (Adabraka, Alajo, Avenor, Odawna, Mataheko, Nima, and Mamobi). The vulnerability of the neighborhoods to flooding is partly due to the low-lying topography (lies a few meters above sea level) and uncontrolled slums (Owusu et al. 2019). The geographical spread of the study sites (Fig. 1) was to ensure that we obtained a wide and representative data.

Primary data for the study were collected using the procedure described in Owusu et al. (2019). In brief, data on impacts of flooding on household livelihoods, mitigations and adaptation strategies, and barriers were collected at household and institutional levels. A combination of stratified and simple random sample survey were conducted in the neighborhoods located in the two residential classes. A total of 320 household heads were interviewed, which comprised 90 and 230 in the second and third class residential areas, respectively. Data collected included background information, experience, and narration of impact of flooding in the area of residence and individual and/or community preparedness and response.

Interviews were conducted with construction engineers and relevant institutions such as the Departments of Physical Planning, Development Planning, Urban Roads, and National Disaster Management Organization (NADMO), Geological Survey Department, and the Ghana Water Company Limited (GWCL) in Greater Accra Metropolitan Assembly (AMA) to provide further insights into the household surveys and preparedness and flood-mitigation measures rolled out by the metropolis to control perennial flooding in Accra. To supplement the data on flood-mitigation measures in the Accra, secondary data collection was done to obtain information on past and current public (government) invested projects to tackle flooding and its associated impacts on socio-economic livelihoods on the population. The secondary

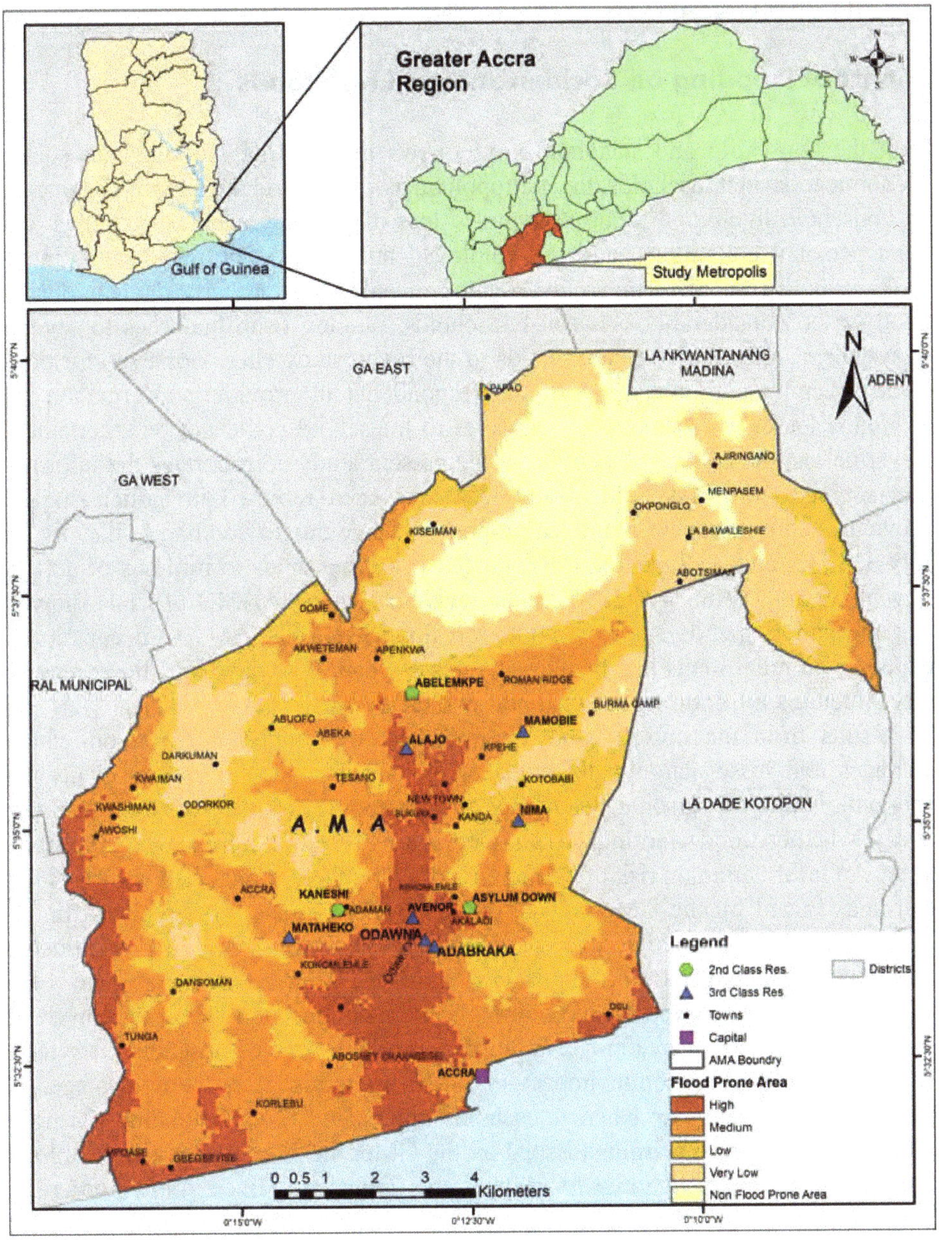

Fig. 1 Map of Accra showing the study locations and flood-prone areas. (Source: Authors)

data were sourced from newspapers, journal articles, and government documents. Both the primary and secondary data were manually analyzed to establish themes and patterns based on the objectives of the study.

Results and Discussion

Impact of Flooding on Socio-economic Livelihoods

Both the household and stakeholder interviews indicate that flooding has socio-economic tolls on households, the metropolitan assembly, and the nation. The impact on household livelihood reported included loss of property such as buildings and other personal belongings, and self-employed businesses. The respondents interviews in both residential areas investigated revealed that property damage during flood varied considerably between households, ranging from hundreds to several thousands of dollars across households in the seven study sites. However, for poor record keeping and reluctance of the respondents in providing information on economic losses, exact monetary damages to household could not be ascertained. Nevertheless, the estimates reported in the present study corroborates the information sourced from stakeholders and what has been reported in earlier studies. Previous studies in Accra estimated property damage due to flooding in the city on July 3, 1995, and on June 13, 1997, ranged from thousands to millions of dollars (Kwarteng 2007; Aboagye 2012). Interviews held with NADMO officials showed that the nation spends a lot of money on flood events in AMA – it constitutes expenses on relief items like blanket, mattresses soap, food, and sometimes temporary structures for flooding affected households.

Results from the present study showed that the magnitude of flood-related damages and losses largely depended on household characteristics such as the economic value of personal belongings of affected households, and more importantly, whether flood warning were issued prior to a flooding episode. The main source of information on risk of flooding reported by the respondents was flood alert warnings issued by the GMet. Thus, the interviews revealed that reported flood-induced damages were high among respondents who own several and/or valuable properties such as electronic appliances, furnished homes, and those who were engaged in self-employed business in the same neighborhood where they lived or in another flood-prone area in Accra. Further, results from the household interviews demonstrated that economic impact of floods in terms of property damage, in general, tends to be low when households anticipated risk of flooding during a rain storm/event from warnings issued by the GMet. In that case they were able to move some of their belongings to relatives and friends who lived in low flood risks areas in the city.

Further, physical destruction of buildings means household loss their homes. Over 65% of the respondents in both residential areas reported that they or someone they knew had been homeless for a few weeks to several months after flooding in the last decade. The respondents mentioned that being homeless has both economic and psychological consequences on their households, though the latter could not be readily quantified. According to a respondent: "...Being homeless, especially when you have a family makes you feel like you have life without soul or sailing without direction."

In addition to property losses, all the respondents emphasized that flooding in Accra posed physical danger on lives and health of households, such as injuries on the day of flooding and an outbreak of cholera and other water-borne diseases after flooding. About 30% of the households, mainly in Alajo, Avenor, Odawna, and Adabraka, reported that they have lost a relative or someone they knew due to flood-related disasters in the past decade, probably due to the close proximity of the neighborhoods to the Odaw River which frequently overflows its banks during rainstorms. Cost of treating injuries immediately or a few days following flood-disasters was reported to be free of charge because AMA took care of the medical bills, whereas cases reported several days after flooding were not treated for free. This means "unless a patient is a national health insurance scheme participant, they have to pay for cost of health care services." The present study demonstrates that the overall impacts of flooding on household livelihood in Accra may be higher than previously thought. According to Maskey and Trambauer (2015), hydro-climatic hazards have tangible impacts that are easily quantifiable and intangible that are not easily quantifiable. Owusu and Teye (2015) found that property loss and death associated with urban flooding in Accra could be remarkably high, but evaluation is limited by lack of data. It is interesting and relevant for future studies to map and quantify the actual socio-economic, health, and psychological burden of flooding on households in Accra using proxy data such as cost of treating injuries and/or disease morbidity, lost wages, time spent for caring for sick or injured household member, etc., to provide a robust documentation and clear picture of impact of floods. Such information can be very useful for prioritizing government- and private-led schemes for supporting flood-affected households in Accra.

Flood Control and Adaptation Measures

Results from the interviews and detailed review of secondary literature demonstrating the detrimental impact of flooding on household livelihood and economic implications for the nation (e.g., Mensah and Ahadzie 2020; Ahmed et al. 2018) suggest that flood control in the short-term or adaptation in the long-terms is imperative. However, it is intriguing to point out that the long flood history of Accra has, indeed, shaped the interaction between local and/or metropolitan activities and development, and flood control and adaptation practices. In the following section, the individual and communities, and metropolitan flood control and adaptation interventions, as well as some pertinent barriers to adaptation, are discussed. To better understand the spatial dimensions of community interventions, that is, whether household flood control and/or adaptation strategies are influenced by their physical local or vulnerability to flooding, flood control, and/or adaptation strategies reported by the respondents are discussed in light of the two residential classes investigated.

Individual and Community

Results indicated that individuals in the study areas are coping by adopting a combination of flood control and adaptation practices. For respondents who lived a few meters away from main drains, a key control measure reported was the digging of temporary channels from their homes to the main drain to ensure free flow of water during torrential rain event. However, the practices, in general, were reported to be less effective and less reliable because the main river channel is often choked.

Eighty-five percent of the respondents in marginal areas like flood plains and swampy areas indicated that they have adapted to flooding by taken drastic accommodation steps. The key reported ones included: (i) changing from building expensive permanent concrete block structures to building cheaper temporary structures such as simple wood structures or metal containers. Fifty-eight percent of the respondents who lived in temporary structures held the opinion that living in those structures were necessary to alleviate huge capital losses when the structure is damaged or lost due either to demolition exercises periodically carried out by AMA to bring down unauthorized structures on and near waterways, or the structure is damaged during flooding. (ii) The respondents in both residential classes reported that they also controlled flooding by desilting drainage channels in their neighborhoods. The cost of dredging is sometimes fully supported by AMA through the local assembly. However, they emphasized that such a support was not regular. Therefore, the common practice has been organizing communal labor to dredge some of the sections of the drainage channels themselves. Further, the interviews revealed a few relatively rich individuals in both residential areas sometimes employed the services of private dredging companies to desilt all or a section of drainage channels closed to their homes to allow free flow of water, which helps prevent flooding. (iii) Another individual adaptation practice reported by 52% and 48% of the respondents in the third and second class residential areas was temporary vacation of homes located on and near waters during the rainy season, especially when the GMet issued an alert that flooding in Accra was imminent. Findings are consistent with previous observations (e.g., Mensah and Ahadzie 2020) showing that in flood-prone areas, people adopt reactive strategies like relocating and protecting properties, for instance, by practicing early transfer of personal belongings to safe places prior to onset of rains. The present study further showed that flood-forecasting by the GMet plays a crucial role in strengthening household seasonal adaptation practices and thus, helps to alleviate the negative impact of flooding on socio-economic livelihoods.

Households in the highly flood-prone residential areas, particularly in Alajo, Adabraka, Avenor, and Asylum Down, have adapted to the economic shocks of flooding by enrolling in life and property insurance schemes, which according to them provided an economic cushion in times of flood-related disasters. However, because premiums and monthly charges can be prohibitively expensive for, especially, low-income households, only a few of the respondents (18%) reported that they had life and/or property insurance.

Metropolitan Assembly/Government

The household and stakeholder interviews pointed out two key areas of institutional or government 'action plans" to managing flooding in Accra. A common approach, which has for many decades been adopted by AMA, was routine dredging of water channels, notably the Odaw River and Korle-Lagoon. The Odaw River has a wide catchment area and covers most parts of Accra and feeds into the Korle-Lagoon, which empties the content of the Odaw and other drainage channels into the Gulf of Guinea (Owusu Boadi and Kuitunen 2002; Ministry of Works and Housing 2019). Table 2 presents an overview of government flood control and adaptation projects in Accra in the last decade implemented during the rainy season before or after flood disasters. Table 2 illustrates two important considerations: (i) the state has spent substantial amount of money in the past decade on measures to prevent flooding in Accra, and (ii) flood prevention measures have mainly been dredging to unblock major waterways and drainage systems prior to the rainy season. Findings of the study highlight that although the dredging projects required massive capital investment, the persistence of flooding in Accra is a strong evidence that the project may be providing ad hoc solutions with very minimal long-term impacts as previously reported by Owusu et al. (2019). Evidence here suggests a need for government to rather consider massive capital investments such as the construction of more robust drain, which can offer sustainable solutions for controlling flooding in Accra and in other flood-prone cities in Ghana.

Another institutional intervention reported was the demolition of unauthorized buildings on or near waterways to allow free flow of water. Demolition of building is an exercise regularly carried out by AMA prior to the start of the rainy season, especially when torrential rains were expected in Accra and its surroundings. The interviews revealed that likewise dredging, demolition also only offers a short-term solution for managing flooding because the demolished structures are re-erected shortly after the AMA-led demolition exercise is completed.

Summary Types of Adaptation Measures to Flooding in Accra

Alcamo and Olesen (2012) proposed that climate change adaptation measures could be classified into the following: autonomous vs. planned, anticipatory vs. reactive, centralized vs. decentralized, and gradual vs. abrupt. Although these classifications were originally proposed in relation to climate change, they could be applied in the present study to classify the adaptation measures to flooding. Results showed that adaptation at the individual level encompasses both autonomous and planned adaptation practices. Decisions taken by households on where to relocate to when flood warnings are issued by GMet are mainly taken at the family level. The planned adaptation measures adopted by the individual included the construction of improvised channels from homes to main drains to ensure free flow of water. Dredging of drainage channels by community and/or government prior to the onset of rain is both anticipatory (i.e., in anticipation to control risk of flooding during the rainy season)

Table 2 Overview of government investments (2010–2020) to control flooding in the Greater Accra Metropolitan Assembly

Project	Date (period)	Description	Cost (US$)	Sponsors	Reference
Greater Accra Resilient and Integrated Development Project	2019–2025	Phase 1: mitigating the flood risk for a 10% annual exceedance probability flood in the Odaw River Basin	200 million	Dutch Government and the World Bank	The World Bank (2019)
		Phase 2: expanding flood mitigation and drainage improvements to additional priority basins (such as Lafa, Sakumo, Kpeshie, and Densu asins)	300 million	Dutch Government and the World Bank	The World Bank (2019)
		Phase 3: scaling up the project to the remaining priority basins of the Greater Accra Region (GAR) to mitigate flood risk in the entire GAR	400 million	Dutch Government and the World Bank	The World Bank (2019)
Korle Lagoon Ecological Restoration Project	2015–2016	Dredged the Odaw drain and Korle Lagoon to solve the perennial flooding in Accra in order to save lives and property	10 million	–	Arthur-Mensah (2017)
Relief and Repair Project	2015	A relief and repair efforts following flooding with an associated fire killed about 200 people in the commercial hub of Accra	12 million	Government of Ghana	The Economist Group (June 23rd 2016)
Accra Sanitation, Sewer, Stormwater and Drainage Alleviation Project	2013	Improving the drainage system in Accra to control the perennial flooding and sanitation challenges in the city	595 million	Loan facilities with the EXIM Bank of the USA and Standard Chartered Bank	BBC (2015)
Dredging the Odaw River	2011	Improving the drainage system of the Odaw	1.2 million	–	The Daily Graphic (2012)
Dredging Committee for Chemu Lagoon	2010	Improving the drainage system of the Chemu	2 million	–	Business Ghana (2010)

and reactive (i.e., to avoid similar flood-related damages and loss experienced during previous floods). Centralized adaptation measures to flooding in Accra encompass AMA or government sponsored projects carried out to dredge principal drainage channels; the commonly reported were the Odaw River and Korle-Lagoon. The decentralized adaptation measures involved dredging minor drainage channels sponsored through community fundraising. Finally, gradual adaptation measures to flooding are evidenced by the individual's decision to change from building more permanent and expensive homes on or near principal waterways to avert high economic losses when the structures are damaged due to flooding or government demolition. Further, as mentioned previously, a few of the respondents have adopted gradual adaptation strategy by enrolling in life and property insurance schemes against accidents and life-threatening disasters. Conversely, AMA- or government-led demolition of structures before onset of rains to reduce risk of flooding and associated damages and causalities is mainly an abrupt adaptation measure.

Limits and Barriers to Adaptation to Flooding

A notable limitation to individual and community funded dredging reported was noncooperation and noncompliance among some community members. It was reported that, often, people are unwilling to contribute financially to enable the community leaders hire the services of private dredging companies. A major reason cited for noncontribution was "to some people this whole dredging activities is the responsibility of the local assembly and the government" [An assembly man at

Fig. 2 (a) Location of residential buildings near open drainage and (b) excavation and disposition of dredged materials to drainage banks, which lead to rapid re-siltation and increases flood risk in Accra

Alajo]. Therefore, there was no need for the individual to take over what the government must do for the community or city.

Further, despite community and/or government massive investments in dredging and related projects, the impact of these projects, in general, seemed to be short-lived because as illustrated in Fig. 2, (i) the open drain system results in faster accumulation of silt and solid waste, (ii) the dredged materials are often left on the banks of the drainage, which leads re-siltation of the waterways after few rain events, (iii) lack of enforcement of town plans and building permits, and (iv) there is also a societal habitat of massive littering and disposing of domestic and industrial waste into drains, leading to rapid refilling of dredged drainage channels. According to Mensah and Ahadzie (2020), improper urban disposal of waste can heighten flood events. Findings here suggest a need for AMA to invest in the construction of more robust drainage systems to build flood resilience in Accra. Interviews with planning and private urban planning consultants emphasized the urgent need for more sustainable development infrastructure in Accra and in other flood-prone cities, for example, by switching from open drain to close drainage system to abate using drainage channels as dumping sites.

Conclusions and Perspectives

Findings illustrate the perennial flooding in Accra that continue to cause massive property damage and occasional loss of lives. The main cause of flooding in the city was found to be the siltation of the Korle-Lagoon and the Odaw River and its tributaries, the construction of houses on waterways, and flood plains due mainly to lax enforcement of building permits. The very nature of development, that is, the building of open drains coupled with poor sanitation attitude among residences, results in the blockage of drains in the city that accounts for most of the flood episodes. This is supported by the fact that the Accra plains where the city of Accra is located is actually the driest part of Ghana with average annual precipitation totally 800 mm yet Accra is the most flooded city in the country.

Findings showed that flooding in Accra has enormous adverse impacts on household income, property, and employment. The main adaptation measures at the level of the individual, community, and government include building less permanent and inexpensive structures to reduce economic cost of flood-induced damages, dredging of drains ahead of rainfall and demolishing of buildings on or near waterways. Nevertheless, the adaptation measures, in general, were reported to be ineffective to make the city resilient to flooding. A major contribution of the study is showing that flood forecasting by GMet plays a crucial role in mitigating the devastating impact of flooding on household socio-economic livelihoods in Accra. Findings also points to the need for city authorities to implement robust measures to make Accra more resilient to flood in the era of climate change. It is suggested that, in the short-term, effort of making sure that dredged materials are not deposited on the banks of the drainage channel to prevent rapid re-siltation is needed. The dredged material could be put into good use in construction and landfill. However, because of high level of pollution, more studies on how safely to dispose of or potentially use the dredged material should be done. Long-term adaptation such as robust engineering solutions, for example, covering open drains and building of retention ponds that will reduce surface runoff in the city, should be considered. There are a few green spaces like the Achimota forest and large undeveloped land near the city that can be created as retention ponds to help solve the perennial flooding problem in Accra. A potential limitation of the study is that the conclusions presented in the present work are based on a sample size of 320 from 10 flood-prone neighborhoods. It is recommended that future similar studies are conducted in other flood-prone neighborhoods in Accra and elsewhere in Ghana to expand the present work to improve the generalization of the findings for spatial planning and interventions to alleviate impact of flooding on urban livelihoods.

References

Aboagye D (2012) Living with familiar hazards: flood experiences and human vulnerability in Accra, Ghana. J Urban Res. https://doi.org/10.4000/articulo.2110

Ahmed F, Moors E, Khan MSA, Warner J, Terwisscha van Scheltinga C (2018) Tipping points in adaptation to urban flooding under climate change and urban growth: the case of the Dhaka megacity. Land Use Policy 79.496–506. https://doi.org/10.1016/j.landusepol.2018.05.051

Alcamo J, Olesen JE (2012) Life in Europe under climate change. Wiley-Blackwell, Chichester

Amoako C (2012) Emerging issues in urban flooding in African cities-the case of Accra, Ghana. Paper presented at the 35th AFSAAP annual conference proceedings 2012

Appeaning Addo K, Appeaning Addo I (2016) Coastal erosion management in Accra: combining local knowledge and empirical research. Jàmbá: J Disaster Risk Stud 8(1). https://doi.org/10.4102/jamba.v8i1.274

Arthur-Mensah G (2017) Government should pay contractors for dredging Odaw-MP. Ghana News Agency. Available at: https://www.modernghana.com/news/780971/government-should-pay-contractors-for-dredging.html. Accessed June 10, 2020

Asumadu-Sarkodie S, Owusu PA, Rufangura P (2015) Impact analysis of flood in Accra, Ghana. Adv Appl Sci Res 6(9):53–78

BBC (2015) Ghana petrol station inferno kills about 150 in Accra. BBC. http://www.bbc.com/news/world-africa-33003673. Accessed 5 June 2015

Business Ghana (2010) Ghana inaugurates dredging committee for Chemu Lagoon, April 1, 2010

CNN World (2001) Ghana floods leave thousands homeless. CNN.com, June 28, 2001

Daily Graphic (2015) Flood disaster profile of Ghana since 1968. Daily Graphic, June 5, 2015

Darko KA (2020) One dead as military, NADMO move to rescue Accra flood victims. 9 June 2020

Dewan TH (2015) Societal impacts and vulnerability to floods in Bangladesh and Nepal. Weather Clim Extremes 7:36–42. https://doi.org/10.1016/j.wace.2014.11.001

Güneralp B, Güneralp I, Liu Y (2015) Changing global patterns of urban exposure to flood and drought hazards. Glob Environ Chang 31:217–225. https://doi.org/10.1016/j.gloenvcha.2015.01.002

IPCC (2000) IPCC special report on emissions scenarios. In: Nebojsa Nakicenovic RS (ed) Special report on emissions scenarios. Cambridge University Press, Cambridge, UK, p 612

IPCC (2007) Climate change 2007: synthesis report. Contribution of Working Groups I, II and III to the fourth assessment report of the Intergovernmental Panel on Climate Change IPCC, Geneva

IPCC (2018) Summary for policymakers. In: Masson-Delmotte V, Zhai P, Pörtner H-O, Roberts D, Skea J, Shukla PR, Pirani A, Moufouma-Okia W, Péan C, Pidcock R, Connors S, Matthews JBR, Chen Y, Zhou X, Gomis MI, Lonnoy E, Maycock T, Tignor M, Waterfield T (eds) Global warming of 1.5 °C. An IPCC special report on the impacts of global warming of 1.5 °C above pre-industrial levels and related global greenhouse gas emission pathways, in the context of strengthening the global response to the threat of climate change, sustainable development, and efforts to eradicate poverty. World Meteorological Organization, Geneva, 32 pp

Jahanbazi M, Egger U (2014) Application and comparison of two different dual drainage models to assess urban flooding. Urban Water J 11(7):584–595. https://doi.org/10.1080/1573062X.2013.871041

Jamshidi O, Asadi A, Kalantari K, Azadi H, Scheffran J (2019) Vulnerability to climate change of smallholder farmers in the Hamadan province, Iran. Clim Risk Manag 23:146–159. https://doi.org/10.1016/j.crm.2018.06.002

Kim Y, Eisenberg DA, Bondank EN, Chester MV, Mascaro G, Underwood BS (2017) Fail-safe and safe-to-fail adaptation: decision-making for urban flooding under climate change. Clim Chang 145(3):397–412. https://doi.org/10.1007/s10584-017-2090-1

Kwarteng EM (2007) Aftermath of floods. Daily Graphic, June, 23, pp 18

Maskey S, Trambauer P (2015) Hydrological modeling for drought assessment. In: Paron P, Di Baldassarre G, Shroder JF (eds) Hydro-meteorological hazards, risks, and disasters, Hazards and disasters series. Elsevier, Amsterdam, pp 263–282

Mensah H, Ahadzie DK (2020) Causes, impacts and coping strategies of floods in Ghana: a systematic review. SN Appl Sci 2(5):792. https://doi.org/10.1007/s42452-020-2548-z

Ministry of Works and Housing (2019) Greater Accra Resilient and Integrated Development Project. The Environmental Impact Assessment [EIA] Study for the Dredging in The Odaw Basin, GARID

Møller-Jensen L, Kofie RY, Yankson PWK (2005) Large-area urban growth observations – a hierarchical kernel approach based on image texture. Geografisk Tidsskrift-Danish J Geogr 105(2):39–47. https://doi.org/10.1080/00167223.2005.10649538

Okyere CY, Yacouba Y, Gilgenbach D (2012) The problem of annual occurrence of floods in Accra: an integration of hydrological, economic and political perspectives. ZEF University of Bonn, Bonn

Owusu Boadi K, Kuitunen M (2002) Urban Waste Pollution in the Korle Lagoon, Accra, Ghana. Environmentalist 22(4):301–309. https://doi.org/10.1023/A:1020706728569

Owusu K, Teye KJ (2015) Supplementing urban water supply with rainwater harvesting in Accra, Ghana. Int J Water Resour Dev 31(4):630–639. https://doi.org/10.1080/07900627.2014.927752

Owusu K, Yankson PWK, Asiedu AB, Tutu-Benefor T (2019) Climate change impacts on two coastal metropolitan areas in Ghana. In: Climate change in Ghana. The human dimension. Afram Publication (Ghana) Limited, Accra

Rain D, Engstrom R, Ludlow C, Antos S (2011) Accra Ghana: a city vulnerable to flooding and drought-induced migration. UN-Habitat, Nairobi

The Daily Graphic (2012) The dead Odaw River – polluted with plastics, garbage & human excreta. GhanaWeb. Mar 13, 2012. http://www.ghanaweb.com/GhanaHomePage/. The Daily Graphic

The Economist Group (June 23rd 2016) One year on, Ghana floods again. Available: http://country.eiu.com/article.aspx?articleid=1994343383&Country=Ghana&topic=Economy_1. Accessed June 15, 2020

The Weekend Globe (2013) Danger: pollution of Odaw increases. http://www.ghanaweb.com/GhanaHomePage/NewsArchive/Danger-Pollution-of-Odaw-increases-286634. Accessed 25 June 2016

The World Bank (2019) Greater Accra Resilient and Integrated Development Project (P164330)

UN-Habitat (2009) Ghana: Accra urban profile. UN-Habitat, Regional and technical Co-operation Division, Nairobi

UN-Habitat (2011) Housing profile. Nairobi

United Nations Office for the Coordination of Humanitarian Affairs (OCHA) (2002) Ghana – floods OCHA, Accra

Viglione A, Rogger M (2015) Flood processes and hazards. In: Paron P, Di Baldassarre G, Shroder JF (eds) Hydro-meteorological hazards, risks, and disasters. Hazards and disasters series. Elsevier, Amsterdam, pp 3–33

Willems P, Arnbjerg-Nielsen K, Olsson J, Nguyen VTV (2012) Climate change impact assessment on urban rainfall extremes and urban drainage: methods and shortcomings. Atmos Res 103:106–118. https://doi.org/10.1016/j.atmosres.2011.04.003

www.macrotrends.net (2020) Accra, Ghana metro area population 1950–2020. https://www.macrotrends.net/cities/21107/accra/population#:~:text=Chart%20and%20table%20of%20population,a%201.58%25%20increase%20from%202019

6

Adaptation of Seaweed Farmers in Zanzibar to the Impacts of Climate Change

Georgia de Jong Cleyndert, Rebecca Newman, Cecile Brugere, Aida Cuni-Sanchez and Robert Marchant

Contents

G. de Jong Cleyndert (✉) · R. Newman · A. Cuni-Sanchez · R. Marchant
York Institute for Tropical Ecosystems, Department of Environment and Geography, University of York, York, North Yorkshire, UK
e-mail: georgiadejongcleyndert@gmail.com

C. Brugere
Soulfish Research and Consultancy, York, UK

Abstract

Seaweed farming is an important alternative livelihood activity that has been heralded as a development success story. It has advanced women's empowerment and economic liberation in coastal communities in Zanzibar, despite recent declines in its production. Using data from 36 semistructured interviews, we explore the impacts of climate change on seaweed farming in Zanzibar and the coping and adaptation strategies available to farmers. Interviews reveal that climatic changes observed in Zanzibar are characterized by increased temperatures, increased winds, and irregular rainfall, and these changes have negatively affected coastal seaweed farming yields and quality. Combined with economic challenges, these environmental stressors are threatening the sustainability of seaweed farming and the wider development impacts that have been gained over the past decades. Establishing seaweed farms in deeper water, using new technologies, could be an adaptation method to overcome rising temperatures; however, there are significant socioeconomic barriers for this to happen. For example, women lack access to boats and the ability to swim. Adaptation options to the increasing impacts of climate change will be possible only with institutional support, significant investment, and through the empowerment of women and the participation local communities.

Keywords

Development · Climate variability · Coastal communities · Gender · Coping strategies

Introduction

The livelihoods of coastal communities are strongly linked to the health of the coastal and marine ecosystems on which they rely (Salafsky and Wollenberg 2000). These socioecological systems are vulnerable to sudden shocks and long-term change, including climate change, and communities often exhibit a high incidence of poverty that can be exacerbated by these shocks (Tobisson 2014; Ferrol-Schulte et al. 2015; Cohen et al. 2016). Alternative and diversification of livelihood activities are popular intervention options aimed at elevating the socio-economic status of coastal communities and reducing the pressure on marine resources (Sievanen et al. 2005). To be successful, however, these alternative livelihood activities must be resilient to fluctuations in ecological, economic, and social systems (Allison and Ellis 2001; Newman et al. 2020).

Seaweed farming is an alternative livelihood activity that has been promoted in many tropical developing countries (Crawford 2002; Sievanen et al. 2005) because of its low initial capital investment and short-farming cycle (Mshigeni 1973; Valderrama et al. 2015) which provides a fast return on investment (Valderrama et al. 2015). The positive socioeconomic impacts of seaweed farming have been documented in countries including the Philippines, Indonesia, Tanzania, and the Pacific Islands (Sievanen et al. 2005; Msuya 2006a; Namudu and Pickering 2006; Arnold 2008). However, these successes have also been contested (Bryceson 2002; Fröcklin et al. 2012) and seaweed farming is prone to environmental and economic boom-and-bust cycles (Valderrama et al. 2015). This raises questions about its resilience to ecological, economic, and social system fluctuations, particularly the impacts of climate change, and therefore about the sustainability of seaweed farming as an alternative livelihood activity (Allison and Ellis 2001).

Case Study of Zanzibar

Zanzibar is a semi-independent archipelago within the United Republic of Tanzania and its population rely heavily on a vulnerable marine resource base (Suckall et al. 2014). Seaweed farming was introduced in 1989 using the off-bottom method to farm *Kappaphycus alvarezii* (commercially known as *cottonii*) and *Eucheuma denticulatum (spinosum)* in the intertidal zone (Fig. 1) (Msuya 2011). This method involves tying algal fronds to ropes attached between wooden pegs driven into the sediment (Eklöf et al. 2012). Farming cycles are 4–10 weeks, depending on growth rates (Eklöf et al. 2012). Seaweed is dried on the ground over several days and sold to a company to be exported and processed into carrageenan (Fig. 1). Seaweed farming spread rapidly throughout Zanzibar and mainland Tanzania and yearly production increased from 800 tons (dry weight) per year in 1990 to about 11,000 t in 2002 (Eklöf et al. 2012). Farming employs 15,000–20,000 people in Zanzibar, of which 90% are women (Msuya 2006a; Fröcklin et al. 2012). In a traditionally conservative Muslim society, seaweed farming provides women with an opportunity to earn a regular cash income, negotiate household needs, and gain economic independence (Wallevik and Jiddawi 2001). It has been contended that seaweed farming has increased women's security at both the household and community level (Wallevik and Jiddawi 2001). However, despite initial successes, the industry in Zanzibar struggles to compete with global seaweed markets and there has been a production decline of 47% between 2002 and 2012 (Eklöf et al. 2012). The combination of low-sale price and seaweed die-offs, linked to climate change and overgrowth of fouling organisms such as epiphytes (Msuya et al. 2007; Msuya 2011; Eklöf et al. 2012), have caused many farmers to reduce their farm size or abandon the activity altogether (Bryceson 2002; Eklöf et al. 2005).

Adaptation initiatives aim to increase the value of seaweed and improve the livelihoods of the women who farm it. For example, the SeaPoWer project has introduced a new technology to farm *cottonii* (the more valuable species) in the

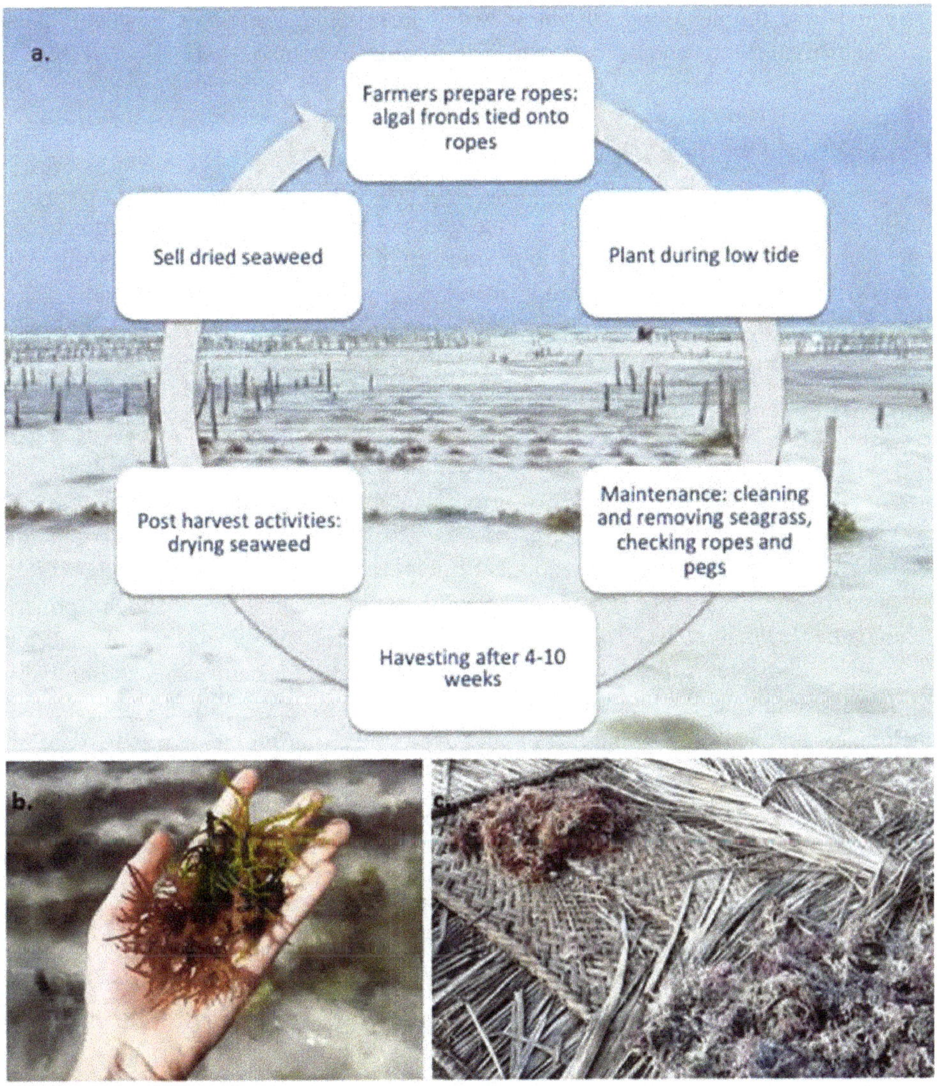

Fig. 1 (**a**) The seaweed farming cycle using the off-bottom method. (Adapted from Fröcklin et al. 2012). Background photo of seaweed farms in Paje, Zanzibar; (**b**) Fresh *Eucheuma denticulatum* (spinosum); (**c**) Two piles of seaweed drying on palm leaf matting outside Jambiani village. (All photos taken by the author)

deeper water (>8 m) using tubular nets and has provided a boat and other equipment to two groups of farmers (Brugere et al. 2019). Value addition training has been provided by academics and NGOs to enable women to process seaweed into more valuable products such as soap, shampoo, cookies, and juice (ZaSCI 2019). Some women have formed "clusters" to share costs of equipment and strengthen their business. Another example of an initiative is *The Seaweed Co.*, a business providing seaweed farm tours and products to tourists, where farmers are employed full time and receive a fixed salary. Adaptation initiatives may increase the sustainability of seaweed farming in Zanzibar; however, their accessibility, effectiveness, and resilience remain to be assessed.

Conceptual Framework

Sustainable livelihoods, adaptation, and gender equality are key concepts to be considered in relation to the unique socioecological system of seaweed farming in Zanzibar. A livelihood is understood in relation to natural, human, economic, and social capital and is considered sustainable if it can cope with present and future shocks and stresses, while not undermining natural resources (Scoones 1998; Serrat 2017; Quandt 2018). Considering the strong interrelations between environmental conditions and seaweed farming outcomes, attention needs to be paid to the implications of environmental change. Where previous literature has emphasized the challenges that environmental change can pose (Hassan and Othman 2019; Makame and Shackleton 2019) more attention needs to be paid to if, and how, people are responding. Therefore, this study adopts principles from the dynamic environmental sustainability of livelihoods (DESL) framework, which focuses on dynamic responses to change (Newman et al. 2020). Typically, those who are unable to cope, by making temporary adjustments, or to make long-term adaptations are vulnerable and unlikely to attain sustainable livelihoods (Scoones 1998). Responses to shocks and stresses include extensification, intensification, and diversification (Suckall et al. 2014). With regard to seaweed farming, these can be understood as increasing the geographical area where seaweed is farmed, increasing the time spent on the farm, and finally, taking on additional livelihood activities. Long-term adaptations are the actions of individuals, communities and governments undertaken for the purpose of improving or protecting livelihoods (Adger et al. 2005). The capacity to adapt is shaped by socio-institutional factors, including social identities and power relations, which include gender inequalities (Brown and Westaway 2011). The sensitivity of seaweed farming to environmental fluctuations and climate change make it particularly susceptible to shocks and stresses (Msuya and Porter 2014) and the unique nature of this female-dominated industry in the context of gender inequalities in Zanzibar, may impact farmers' ability to adapt. Assessing the viability of coping and long-term adaptation strategies is essential for understanding its future importance as an alternative livelihood activity. As such, we focus on adaptive strategies alongside barriers to adoption whilst critically reflecting on how such barriers might be overcome.

Aims and Objectives

This study aims to explore whether seaweed farmers in Zanzibar can adapt to climate change to ensure its continued sustainability as a livelihood diversification option. Firstly, we assess the environmental changes that are occurring in Zanzibar and farmers' perceptions of these changes. Secondly, we identify the challenges that women are facing, particularly in relation to climate change, and the effects these challenges have on production. Finally, we explore how farmers are responding to change, the adaptation strategies available and the barriers to adaptation. In doing so, findings from this work contribute to understanding the effectiveness of adaptation options and to developing recommendations for enhancing the sustainability of seaweed-related interventions.

Methods

Meetings took place with key NGOs and academics working with seaweed farming in Zanzibar to contextualize the project and gather additional background information, found in the supplementary information (S1: List of Academics and NGOs Contacted During Scoping Phase). Thirty-six semistructured interviews were carried out with seaweed farmers from five villages on Unguja, the largest island of the Zanzibar archipelago (Fig. 2, S2: Participant Profiles); participants were sought with the support of academics and NGOs working in Zanzibar. Thirty-four farmers were female, two were male and ages ranged between 23 and 70, with the average age of 45. Four different groups of farmers were selected to represent different types of seaweed farming options on the island (Table 1). Groups were identified during the scoping phase and a purposive sampling strategy was used to gain representation of the four identified groups of seaweed farmers. This number of participants was deemed an adequate sample size to permit case-orientated analysis while providing a new understanding of experience (Sandelowski 1995).

Interviews were semistructured to allow for supplementary information to be incorporated into the discussion. Interviews were carried out in participants' homes with the assistance of a translator and were recorded using a mobile phone if consent was given. The translator had an academic background in marine sciences and translated from Swahili to English. A meeting was held with the translator before interviews commenced to ensure that translation and interpretation of questions were accurate. Furthermore, to minimize misinterpretation, there was mutual consultation between the translator and the researcher throughout translation during interviews to fully unravel answers (Temple and Young 2004). Six pilot interviews were carried out and questions were adapted to make them clearer for participants and to make the challenges section less restrictive and allow participants to talk more openly and in depth about the challenges they were experiencing.

Questions were developed using themes that were based around the key aims and objectives (S3: Interview Questions Used as a Guideline for the Semi-structured Interviews). The first section gathered background information to gain understanding about whether farming contributes to financial stability in the household. The second section asked farmers to consider challenges, how these have changed over time and their coping strategies. Adaptation methods were considered by opening up conversation about deepwater farming and value addition (if applicable) and ways to make farming easier. The final question was very open-ended, allowing participants to freely add any information that they deemed important. Interviews lasted between 20 and 40 min. Participants received a small remuneration for their time according to local customs (4000 Tsh / US$ 1.5).

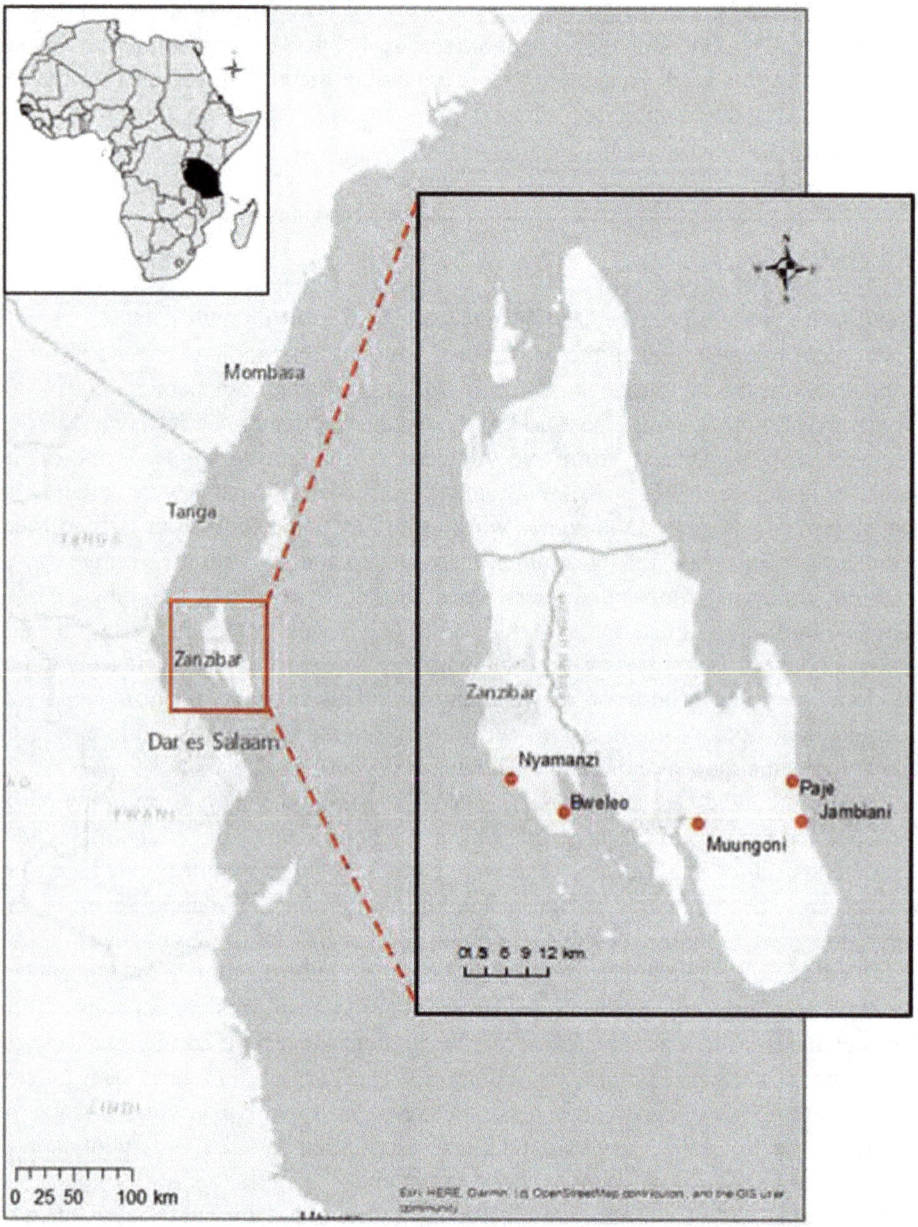

Fig. 2 Map of Zanzibar. Red dots indicate villages where participants were interviewed

Analysis

Interview notes were studied to identify key themes and concepts emerging from the data (Spencer et al. 2003). Thematic categories were based on the objectives but labeled using language from participants to ensure that analysis remained embedded in the data (Spencer et al. 2003). Interview scripts were then systematically coded

Table 1 Summary of the groups interviewed. (No. = number of participants)

Group	No.	Description	Method of farming	Value addition?
Independent Farmers	9	Farmers working alone, receiving no institutional support	Off-bottom	✗
Value addition	11	Farmers making value addition products, part of a "cluster"	Off-bottom	✓
Sea PoWer	10	Received equipment and training to farm in the deepwater using tubular nets	Off-bottom and deepwater	✗
Seaweed Co.	6	Employed 6 days per week, receive a wage. Business sells tours to tourists	Off-bottom	✓

under thematic categories using NVivo 12 (version 12.4.0) (S4: Coding Strategy Used to Analyses the Seaweed Farming Interview Data). Coded information about challenges was organized into a presence-absence data format (S5: Presence Absence Data for Challenges to the Off Bottom Method, S6: Presence Absence of Challenges for Deepwater and off Bottom Methods). Statistical analyses were carried out using R (version 3.6.0). Fisher's Exact tests were run to compare challenges across the deepwater farming ($n = 10$) and the off-bottom method ($n = 36$). Further analysis took place by examining recordings and full interview notes to capture detailed illustrative quotes. Quotes were selected to capture the breadth of challenges and emerging themes.

Results

Seaweed farming is often seen as one of the few options available for women to earn an income. With limited options for other employment, it is an important livelihood activity to alleviate poverty. Income from seaweed farming varied dramatically, ranging from 20,000 to 160,000 Tsh (US$ 9.00–69.00) per month. Farmers noted the considerably large differences in income when the harvest was "good" or "low." Challenges to seaweed farming are numerous and farmers indicated that farming has been affected by changes in climatic factors and nonclimatic stressors over the last 20 years. Climatic variables include increased sea temperatures, increased winds, and irregular rainfall (particularly impacting on the ability to dry the harvested seaweed). Other stressors include low market price and health repercussions, such as back pain and skin irritation. The more valuable species, *cottonii,* cannot be farmed in many areas in Zanzibar due to poor growth and die-offs driven by diseases, such as "ice–ice" disease. As a result of challenges, there has been a huge reduction in the number of farmers. For example, the number of farmers in Bweleo has been reduced from 200 to 60 (Seaweed farmer, VA, Bweleo, July 2019). This reduction in farmers echoes a decline in production across Zanzibar (Fig. 3).

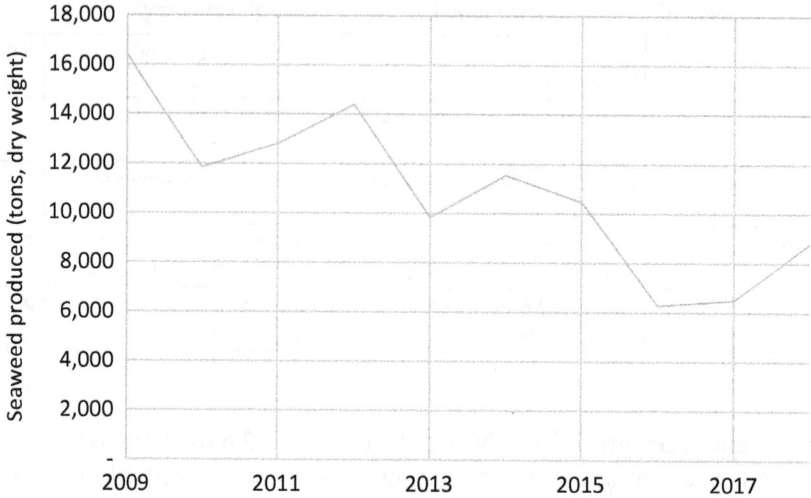

Fig. 3 Seaweed produced in Zanzibar 2009–2018. (Data acquired from Department of Fisheries Development (2019))

Meteorological Evidence

See Fig. 4.

Perceived Climatic Changes and Reported Impacts on Seaweed

Although some seaweed farmers did not observe climatic change or were unaware of why seaweed was not growing well, many were aware of changes and reported that stressors were having a big impact on seaweed farming, reducing yields. The more valuable species *cottonii* cannot be farmed anymore so farmers have to produce the lower value *spinosum*. Table 2 indicates the effects that climate stressors have on seaweed farming.

Adaptation Strategies and Constraints

Although some farmers could see no solution to the challenges they faced, there are some adaptation strategies to mitigate the effect of low market price and climatic stressors (Table 3). One coping strategy is to tend to farms more often, using "more energy" to farm. Value addition is an adaptation to the low sale price of seaweed because it can earn more money. However, it does not provide a solution to the problem of a low harvest.

> Due to the disease, we try to farm seaweed, but we harvest nothing. If we want to make value addition, we have to buy from other farmers. The solution is to get a boat to farm into the deep water. (Female seaweed farmer partaking in value addition, Bweleo, July 2019)

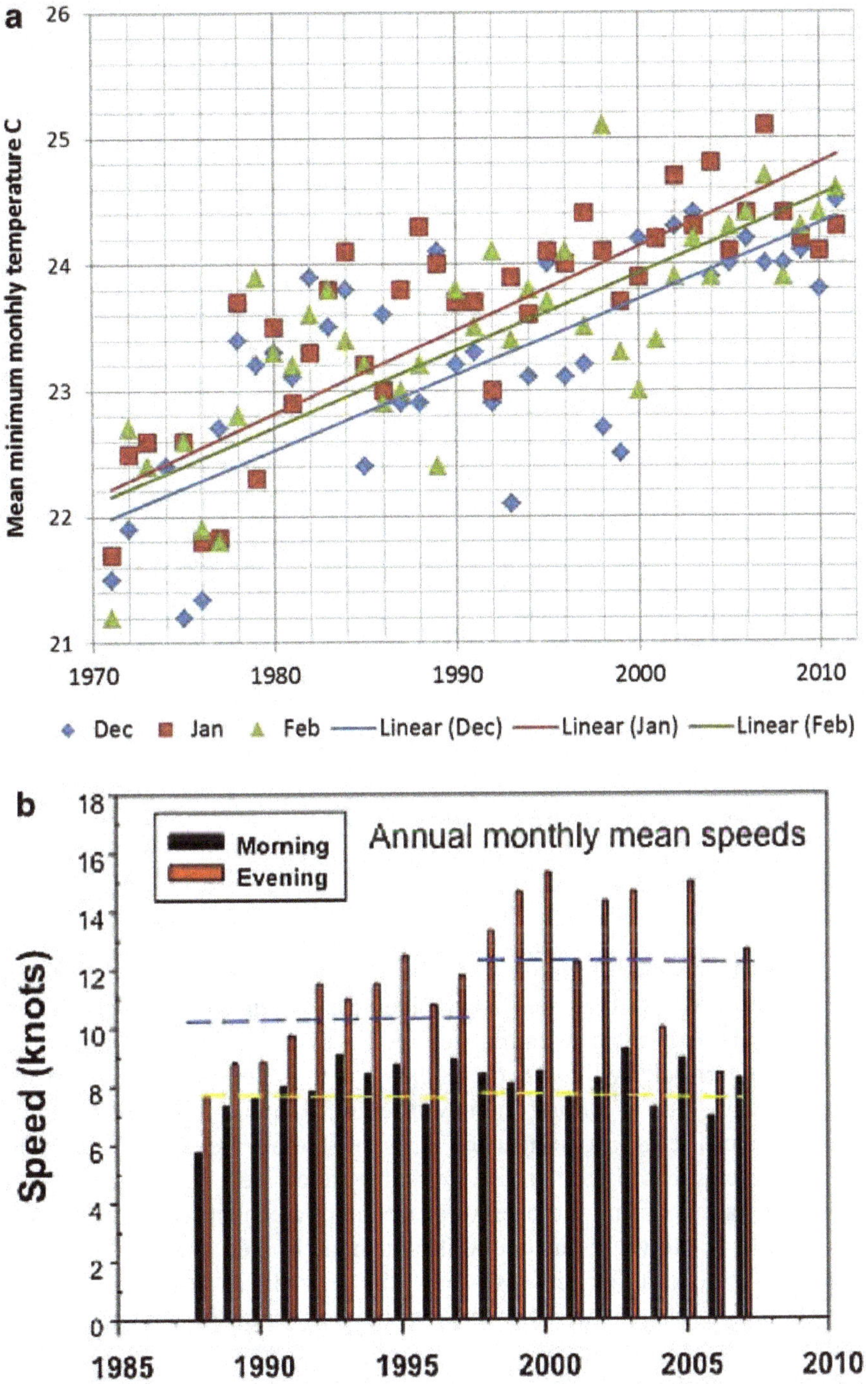

Fig. 4 (**a**) Mean monthly minimum temperature in January and February on Unguja show temperatures have increased strongly over the last 40 years. (Figure from Paul Watkiss, copyright permissions obtained); (**b**) Trends in annual monthly mean wind speeds for Unguja show that wind speeds have increased in recent decades.

Table 2 Effects of climatic stressors on seaweed farming as mentioned by participants

Stressor	Effects	% participants	Example
Increased sea temperatures	More disease – "ice–ice" disease which turns the seaweed white, causes it to rot	61%	*There is disease during the summer, the water got hot and boiled the seaweed* (Independent female farmer, Paje, July 2017) *It is getting worse because now the sun is very high and hot and so the disease is worse* (Female seaweed farmer partaking in value addition, Paje, July 2019)
Increased sea temperatures	Epiphytic algae, causes seaweed to rot	28%	*There is a type of seagrass that grows on the seaweed and causes it to rot... I wanted to cry because there was so much* (Female farmer employed at the seaweed business, Paje, July 2019)
Increased wind speed	Lower yields as seaweed breaks off the rope and farms get destroyed	50%	*Due to the changing of the weather, the seaweed that we plant gets ripped off by the wind. We plant a lot but when we go to harvest it there is not much there* (Independent female farmer, Bweleo, July 2017)
Changes in rainfall patterns	Lower yields as destroys the harvest if it gets wet whilst drying	19%	*During the rainfall we plant it but we don't harvest it because it is difficult to dry it* (Independent female farmer, Paje, July 2017)

Deepwater farming is a group activity that uses a new technology to farm in the deeper and cooler water. It requires the involvement of men to drive a boat and swim to place nets, although some swimming lessons have been provided to some women. SeaPoWer farmers believed that seaweed grows better in deeper water, particularly the more valuable species *cottonii* and it was believed that the new method addresses environmental challenges. A comparison of the off-bottom method and the deepwater method reveals that disease and seagrass infestation were significantly less of a challenge for farmers using the deepwater technology compared to the off-bottom method (Table 4). However, the deepwater method elevated alternative challenges including the presence of grazers (fish-eating the seaweed) and the need for training (Table 4). Farmers have placed fish traps at the bottom of the tubular nets in the deepwater to capture the herbivorous fish to sell. Interviews reveal that there are constraints to adaptation methods (Table 3). For example, independent farmers expressed a desire to partake in value addition but noted that they lacked training and equipment. Similarly, to farm in the deeper water using the tubular technology, farmers need to learn to swim and dive, and have access to a boat. Currently, farmers have to rely on men because boat skills are highly gendered activities that are

Table 3 Adaptation strategies to seaweed farming and associated constraints as mentioned by participants

Adaptation	% of participants	Constraint	Example
Tend to farms more often	19%	Takes more time and energy	*If there is extra wind, I spend more time because I need to check if the pegs and rope are ok* (Female seaweed farmer partaking in value addition, Paje, July 2019) *The main challenge is that I use more energy but get less money* (Female seaweed farmer partaking in value addition, Paje, July 2019)
Value Addition	50%	Seaweed will not grow	*Due to the disease, we try to farm seaweed but we harvest nothing. If we want to make value addition, we have to buy from other farmers* (Female seaweed farmer partaking in value addition, Bweleo, July 2019)
		Require training	*I would like to make products but I need education on how to do it* (Independent female farmer, Paje, July 2019)
		Require equipment	*I have had the training but I can't afford the equipment so I can't do it* (Female SeaPoWer farmer, Dimani, July 2019)
		Takes too much time	*I used to do it in a cluster but we stopped because of time* (Independent female farmer, Paje, July 2019)
Deepwater farming	39%	Ability to swim	*We would love to know how to swim so that we can participate in planting. We need training* (Female SeaPoWer farmer, Dimani, July 2019)
		Ability to dive	*Men should be engaged a lot more with seaweed farming. I think it is difficult for the women to dive. I don't think women will be able to dive* (Male SeaPoWer farmer, Dimani, July 2019)
		Training to drive the boat	*I want to learn to drive the boat myself and not rely on men* (Female SeaPoWer farmer, Muungoni, July 2019)
		Access to a boat	*If we had a boat we would be able to farm in the deeper water* (Independent female farmer, Bweleo, July 2019)
		Grazers	*The fish eat the seaweed. We put traps at the bottom but the fish still come* (Male SeaPoWer farmer, Dimani, July 2019)

Table 4 Comparison of the challenges mentioned by participants for seaweed farming using the off-bottom method and deepwater farming. ***P*** value represents the significance levels from fisher's exact test. * denotes significant results

Challenge	Off-bottom method	Deepwater	*P* value
Disease	64%	20%	$p = 0.028$*
Epiphytes	39%	0%	$p = 0.020$*
Winds	47%	40%	$p > 0.05$
Grazers	6%	50%	$p = 0.0031$*
Training	0%	60%	$p = 0.0000022$*

associated with fishing activity. However, there is strong consensus among the female participants that women would like training to be able to do these activities themselves. These constraints highlight clear socioeconomic barriers to adaptation options.

Discussion

Seaweed farming, using the off-bottom method, is susceptible to shocks and stresses and climatic changes are clearly having a big impact on seaweed farmers. Initial benefits associated with seaweed farming, such as increased household income and job opportunities, become less obvious as harvests are increasingly unreliable which leads to greater insecurity. Many farmers observed increases in wind speed and temperature, which is in alignment with meteorological data and other studies (Hassan and Othman 2019; Makame and Shackleton 2019). The presence of "ice–ice" disease, which causes a discoloration of the seaweed thali and affects the quality of seaweed, is linked to changes in light intensity and temperature (Largo et al. 1995). The more valuable species (*Cottonii)* is particularly sensitive to environmental fluctuations, and seaweed die-offs caused by "ice–ice" disease is a widespread issue and has been long documented in Zanzibar (Msuya et al. 2014; Msuya and Porter 2014). Moreover, the invasion of epiphytic algae, which causes the seaweed to rot (Critchley et al. 2004; Vairappan 2006), has also been linked to increased variability in water temperatures (Tsiresy et al. 2016). The emotional response of farmers during the interviews to this problem clearly indicates the effect it has on income and human capital because of the economic insecurity and emotional response it causes. The difficulty of high die-off rates and therefore low yields is compounded by irregular rainfall patterns, which cause post-harvest loss of yields if it rains during the drying process, and a low-sale price. The combination of environmental and economic challenges results in a very low income for farmers and therefore threatens the economic capital of farmers by reducing their ability to generate a stable income. Seaweed farming is clearly susceptible to economic and environmental shocks and stresses. Given that future projections estimate that temperatures will increase by 1.5–2 °C by the 2050s (Revolutionary Government of Zanzibar 2013), seaweed die-offs will be exacerbated and its sustainability is therefore questionable if adaptation is not achieved. Other challenges, such as poor health effects (Fröcklin et al. 2012), further impact sustainability by negatively affected human capital. The off-bottom method is negatively impacting human and economic capital and is becoming increasingly vulnerable to shocks and stresses. Therefore, it is important to understand indigenous perceptions about climate change and its impacts to assess the dynamic responses to changes and to determine suitable adaptation strategies to attain sustainable livelihoods.

There are a number of short-term and long-term adaptation strategies available to farmers. Intensification of farming (i.e., using "more energy" to tend to farms) is a short-term coping strategy in response to low production, high winds, and low sale price. Value addition is a form of long-term diversification, aiming to address the issue of a low sale price that farmers get as a result of their weak bargaining power with seaweed buyers. Interviews reveal that women clearly see the benefit of value addition through an increase in their income. Moreover, deepwater farming can be seen as either a form of extensification or a migration to a new production environment that successfully addresses many environmental challenges and farmers believe that they can earn more money by farming the more valuable species. Although there are still uncertainties regarding the outputs of these long-term adaptation strategies (value addition and deepwater farming), and further monitoring will be required to assess the impact that this has on the farmers' household income, the attitudes of farmers engaging with these activities are positive. The farming innovations are also bringing about additional benefits such as greater social capital by empowering women producers and elevating women's status in society (Brugere et al. 2020).

However, successful adaptation requires an enabling environment dependent on environmental, economic, social, and institutional factors and some strategies are more effective that others, particularly in relation to climate change. Although some long-term adaptation strategies are having positive impacts, there are a number of barriers to adaptation that warrant further attention, and adaptive strategies must be analyzed in the context of these barriers to assess their effectiveness. Given that climatic change is a major challenge for seaweed farmers that is likely to be exacerbated in the future, many adaptation strategies will not be adequate in ensuring the long-term sustainability of farming. For example, intensification of farming and value addition activities will not be resistant to climatic stressors because they do not address the inability to grow seaweed. Moreover, there are economic considerations that may inhibit farmers' ability to adapt; unlike the off-bottom method, deepwater farming and value addition both require large initial investment to purchase equipment. If farmers do not have the financial capacity to make this initial investment, adaptation will not be possible unless enabled through institutional or NGO support (Wright et al. 2014).

Social factors will also either inhibit or enable adaptation. Deepwater farming and value addition both require substantial training and knowledge sharing. Many independent farmers interviewed were aware of adaptation methods but were unable to access them, highlighting the importance of social collaboration as a critical enabling environment to promote knowledge sharing for adaptation (Fu et al. 2011). Moreover, deeply engrained social practices and the complex nature of gender biases and power relations will be significant barriers to overcome for adaption methods. Labor-led intensification is characterized by an increased time burden, which disproportionately affects women due to their roles within the household and as child-carers (Wodon and Blackden 2006). Therefore, it is unlikely to be sustainable because of the impact on human and social capital by reducing time available for other roles that are typically dominated by women. The complex nature of gender biases and power relations also significantly affects deepwater farming, which currently entails the involvement of men. Gender power dynamics must be monitored to maintain (and promote) the positive impact to date of seaweed farming on women's security at the household and community level (Wallevik and Jiddawi 2001). Interviews revealed women want to learn to swim and drive the boat to reduce their dependency on men, highlighting a shift in traditional attitudes (Brugere et al. 2019; Brugere et al. 2020). Deepwater farming represents a shift in gender attitudes because it challenges the traditional belief that the deepwater is an area accessed by men due to women's limited mobility and role in society (Fröcklin et al. 2014). The government is promoting gender balance by increasing the number of female village leaders and females in government offices which shows a wider change in gendered roles. However, despite signs of attitude shifts, deeply engrained cultural practices and ways of thinking, held by both men and women, require repeated action, support, and perseverance over extensive periods of time to evolve (Brugere et al. 2019).

Lastly, government involvement will impact the success of adaptation methods by providing (or inhibiting) an institutional enabling environment. Policies and institutions play a major contributing role to the sustainability of coping and adaptation strategies (Osman-Elasha et al. 2006). Governance mechanisms aimed at adaptation can support coping strategies by providing training, technical support, and financial support (Jabeen et al. 2010). It is promising that there is government attention on seaweed farming in Zanzibar, highlighted by the recent appointment of National Seaweed Day to emphasize the importance of the activity and previous governmental attempts to increase the price of seaweed (Davis 2011). Moreover, there are currently plans to construct a processing plant on Pemba which is a large-scale value addition project (IPPmedia 2019). However, the government should be investing in technologies that will be resilient to climate change, not only promoting value addition activities which will be an ineffective coping strategy in the long-term.

Overall, short-term coping strategies are often ineffective and have resulted in many people ceasing to farm as seen by the reduction in farmer numbers in Bweleo and the reduction in production. Although long-term adaptation methods provide promising ways to increase the sustainability of seaweed farming by overcoming economic and environmental challenges, particularly those relating to climate change, there are significant socioeconomic barriers that need to be overcome. This will only be achieved through a supportive enabling environment with participation of local communities and institutional support (Sietz et al. 2011).

Future Study

Although 36 participants was deemed sufficient to enable thorough analysis due to the recurrence of themes during interviews, the study would benefit from incorporating the experiences of farmers on the other islands in Zanzibar, particularly Pemba where seaweed farming is a particularly important livelihood activity. More investigation is required into the potential cobenefit of using fish traps at deepwater farms to see if the catch of fish can outweigh the loss of seaweed yield. Moreover, there should be more research into drying techniques in order to reduce the post-harvest loss due to irregular rainfall, or the possibility of farmers selling seaweed fresh opposed to dried. Additionally, the complex issue of gender dynamics requires further study, particularly the possible shift in power as a result of extensification into deeper water, which could have major impacts on the long-term sustainability of seaweed farming as a way of empowering women.

Wider Implications

Seaweed farming is being promoted as an alternative livelihood activity in many tropical developing counties, including Zanzibar, Indonesia, and Philippines. However, it is important that alternative livelihood activities are resilient to ecological and social system fluctuations, particularly climate change (Allison and Ellis 2001). In order to maintain yields, efforts need to focus on seaweed farming adaptation strategies that will be resilient to climate change. Deepwater farming using the new tubular technology shows the most promising adaption method to environmental challenges. However, it requires significant investment and training to ensure its success. Interestingly, despite negative impacts of climate change on the growth of seaweed that have been reported in Zanzibar and elsewhere, seaweed aquaculture is gaining recognition as a climate-change mitigation strategy by its ability to act as a carbon sink (Duarte et al. 2017). Although it would be minor at this scale, it is worth exploring how climate change mitigation policies that provide economic compensation for the environmental benefits brought about by seaweed farming could help investment and could generate a new market for seaweed production (Duarte et al. 2017).

Conclusion

Seaweed farming is still heralded as a success story and is responsible for women's empowerment and economic liberation. However, climatic stressors such as increased sea temperatures, high winds, and variable rainfall reduce seaweed growth and quality. Given the current environmental and socio-economic challenges, seaweed farming provides an unreliable income. Despite the cultural importance as a livelihood activity, the future of seaweed farming is uncertain, particularly as the impacts around climate change are likely to increase. Individual short-term coping strategies, such as intensification of tending to seaweed farms are unlikely to be effective in the long-term. The sustainability of seaweed farming is reliant on long-term adaptation methods that will require adopting new technologies, overcoming significant socioeconomic barriers, and will demand substantial institutional support. It is important to support adaptation strategies that are codesigned with communities, and that use holistic approaches that embrace the technological, individual (social and economic), and institutional dimensions of climate change adaptation.

Supplementary Information

S1: List of Academics and NGOs Contacted During Scoping Phase

Name	Organisation
Narriman Jiddawi	Department of Fisheries Development
Flower Msuya	Zanzibar Seaweed Cluster Initiative
Alice Mushi	Milele Foundation
Cecile Brugere	SeaPoWer
N/A	The Seaweed Company

S2: Participant Profiles

	Participant	Group	Age	Gender	Village	Status	Children	Electricity	Water
1	SWC1	SWC	48	F	Paje	Married	3	1	1
2	SWC2	SWC	33	F	Jambiani	Divorced	5	0	0
3	SWC3	SWC	66	F	Paje	Married	8	0	1
4	SWC4	SWC	44	F	Paje	Married	2	0	1
5	SWC5	SWC	25	F	Paje	Divorced	3	1	1
6	SWC6	SWC	23	F	Paje	Married	0	1	0
7	IND1	IND	34	F	Paje	Divorced	4	1	1
8	IND2	IND	54	F	Paje	Married	8	0	1
9	IND3	IND	70	F	Paje	Divorced	1	0	1
10	IND6	IND	58	F	Paje	Divorced	4	0	0
11	IND7	IND	57	F	Paje	Widowed	3	0	0
12	IND8	IND	39	F	Paje	Widowed	5	0	1
13	IND9	IND	32	F	Paje	Married	4	0	1
14	IND10	IND	42	F	Paje	Divorced	3	1	1
15	IND	IND	51	F	Bweleo	Widowed	6	1	0
16	SP1	SP	64	F	Nyamanzi	Married	5	1	0
17	SP2	SP	46	F	Dimani	Married	6	0	1
18	SP3	SP	48	F	Dimani	Married	12	0	0
19	SP4	SP	40	F	Nyamanzi	Married	3	0	0
20	SP5	SP	41	M	Dimani	Married	3	0	0
21	SP6	SP	38	F	Muungoni	Married	7	1	0
22	SP7	SP	52	F	Muungoni	Married	5	1	0
23	SP8	SP	43	F	Muungoni	Married	2	0	0
24	SP9	SP	52	F	Muungoni	Married	7	1	0
25	SP10	SP	24	F	Muungoni	Married	2	0	0
26	VA1	VA	40	F	Paje	Married	4	1	1
27	VA2	VA	48	F	Paje	Married	3	0	1
28	VA3	VA	46	F	Paje	Married	2	1	1
29	VA4	VA	50	F	Paje	Married	4	1	1

(continued)

	Participant	Group	Age	Gender	Village	Status	Children	Electricity	Water
30	VA5	VA	42	F	Paje	Married	3	1	1
31	VA6	VA	54	F	Bweleo	Married	6	1	0
32	VA7	VA	25	M	Bweleo	Single	0	1	0
33	VA8	VA	45	F	Bweleo	Married	4	1	0
34	VA9	VA	56	F	Bweleo	Married	6	1	0
35	VA10	VA	40	F	Paje	Married	3	0	1
36	Va11	VA	50	F	Paje	Married	6	0	0

S3: Interview Questions Used as a Guideline for the Semi-structured Interviews

Registration:

Age:
Gender:
Village/District:
Single/Married/Widowed:
Number of children:
Electricity in home:
Water piped in the home:

Questions

Background

How long have you been farming seaweed?
Why did you decide to start farming seaweed?
What do you think are the main benefits coming from seaweed farming?
What type of method do you use (shallow water/deepwater)?

- If you do both, why? If you changed from the traditional method to the new method, why?
- When did you start farming the new method?

What type of seaweed do you farm? *Cotonii* or *Spinosum?*
How much time do you spend farming seaweed in the traditional/deepwater method?
How many days a week do you spend farming?

Economic

Who do you sell the seaweed to?
How much money do you make from seaweed farming per month?
What is the price per unit (kg bag of dried seaweed) Are you able to negotiate the prices?
Which method earns more money? Which species earns more money?
Has the value of the seaweed you farm increased, decreased or stayed the same in the last 20 years?

How essential is seaweed farming to the income of the household?

Do you have any other sources of income other than seaweed farming?

What is the money from seaweed farming spent on?

Do you have a say on how the money from seaweed farming is spent?

What is the cost of the off bottom culture equipment per cycle of production?

If you have farmed using both methods, how do you think the deepwater technique compares with the traditional technique, in terms of

- Seaweed seedlings [more expensive, less expensive, the same]
- Equipment (ropes, pegs versus nets, strings, PVC tubes, boat, petrol…)
- Time/labour spent farming (more, less, the same)

What challenges are you facing? (Off bottom method)

Does the seaweed grow well?

Are these new/emerging challenges or have they always existed?

Are these challenges increasing/decreasing or staying the same?

Do the challenges vary across different seasons?

How do you respond to/cope with these challenges?

What would help to stop these problems?

Are you facing any health repercussions?

- Have these increased with exposure or stayed the same?
- How do you cope with or manage these health implications?

What does your family think of farming the new method/the old method?

- Does the time taken to farm seaweed affect your family?

Has there been any spatial conflict with other users of the ocean space? – eg does tourism or fisherman prevent you from farming?

- How do you manage this?

Is there anything that would make it easier for you?

Do you farm in the deepwater?

How much time do you spend on this?

What effect does this have on your income?

What are the challenges?

Do you do any value addition – making products?

How much time do you spend on making products? What equipment and/or inputs are you using?

How/where do you sell your products?

What products do you make?

Is there anything else that you would like to add or say?

S4: Coding Strategy Used to Analyses the Seaweed Farming Interview Data

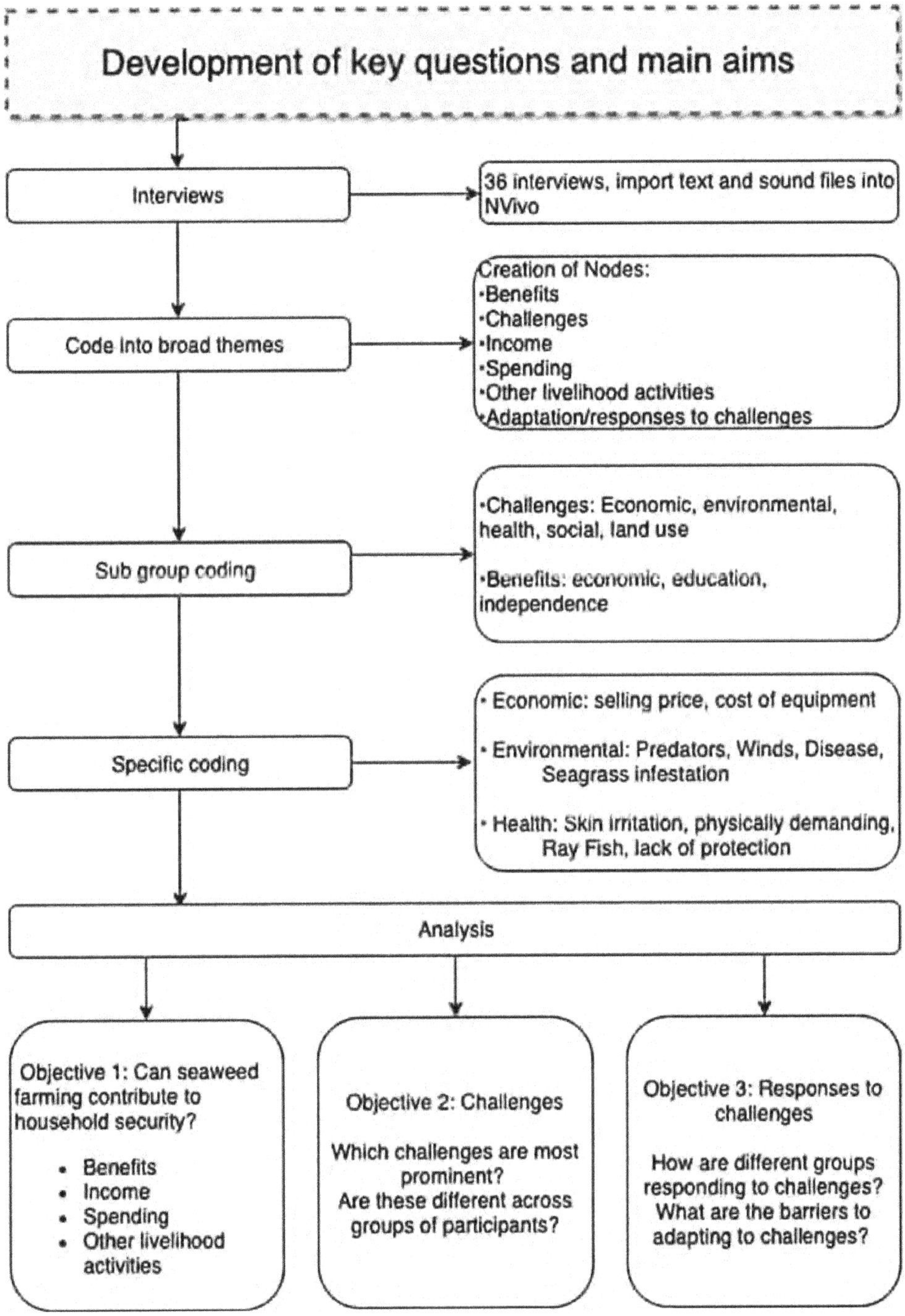

S5: Presen6ce Absence Data for Challenges to the Off Bottom Method

Group	Method	Price	Equipment	Disease	Seagrass	Climate	Winds	Grazers	Health	Training	SpatialConflict	Dryin
SWC	Off_Bottom	0	0	1	0	1	0	0	1	0	0	0
SWC	Off_Bottom	1	0	1	1	0	0	0	0	0	0	0
SWC	Off_Bottom	1	0	0	1	0	0	0	1	0	0	0
SWC	Off_Bottom	1	0	1	0	1	1	0	0	0	0	0
SWC	Off_Bottom	0	0	0	0	0	1	0	0	0	1	0
SWC	Off_Bottom	0	0	0	1	0	1	0	0	0	0	1
IND	Off_Bottom	1	1	0	1	1	1	0	1	0	1	1
IND	Off_Bottom	1	0	1	0	1	1	0	1	0	0	1
IND	Off_Bottom	1	1	0	1	0	1	0	0	0	0	0
IND	Off_Bottom	1	0	0	0	1	1	1	1	0	1	1
IND	Off_Bottom	1	0	1	0	0	1	0	1	0	1	0
IND	Off_Bottom	1	1	1	0	1	0	0	0	0	1	0
IND	Off_Bottom	1	0	0	0	0	0	0	0	0	1	1
IND	Off_Bottom	1	1	1	0	0	0	0	1	0	0	1
IND	Off_Bottom	0	0	0	1	0	1	1	0	0	0	0
SP	Off_Bottom	1	1	1	1	1	1	0	0	0	0	0
SP	Off_Bottom	1	1	1	1	1	0	0	1	0	0	0
SP	Off_Bottom	1	1	0	1	0	1	0	1	0	1	0
SP	Off_Bottom	0	1	1	1	1	0	0	1	0	0	0
SP	Off_Bottom	0	0	1	0	1	1	0	1	0	0	1
SP	Off_Bottom	1	1	1	0	1	0	0	0	0	1	1
SP	Off_Bottom	0	1	1	0	0	0	0	0	0	1	0
SP	Off_Bottom	0	1	1	0	0	0	0	1	0	1	0
SP	Off_Bottom	0	1	1	0	1	0	0	0	0	0	0
SP	Off_Bottom	0	0	1	0	1	1	0	0	0	1	0
VA	Off_Bottom	1	0	1	0	1	0	0	1	0	1	0
VA	Off_Bottom	1	0	0	1	0	1	0	1	0	0	0
VA	Off_Bottom	0	0	0	0	1	1	0	0	0	1	1
VA	Off_Bottom	1	0	1	0	1	0	0	0	0	1	0
VA	Off_Bottom	1	1	1	0	1	0	0	1	0	0	0
VA	Off_Bottom	0	1	1	0	0	0	0	0	0	0	0
VA	Off_Bottom	1	0	0	1	0	0	0	1	0	1	0
VA	Off_Bottom	0	0	0	1	0	0	0	1	0	0	0
VA	Off_Bottom	1	1	1	0	0	0	0	0	0	0	0
VA	Off_Bottom	0	1	1	1	1	1	0	1	0	1	0
VA	Off_Bottom	0	1	1	0	1	1	0	1	0	1	1

S6: Presence Absence of Challenges for Deepwater and off Bottom Methods

Group	Method	Equipment	Disease	Seagrass	Climate	Winds	Grazers	Training	SpatialConflict	Health
SWC	Off_Bottom	0	1	0	1	0	0	0	0	0
SWC	Off_Bottom	0	1	1	0	0	0	0	0	0
SWC	Off_Bottom	0	0	1	0	0	0	0	0	1
SWC	Off_Bottom	0	1	0	1	1	0	0	0	0
SWC	Off_Bottom	0	0	0	0	1	0	0	1	0
SWC	Off_Bottom	0	0	1	0	1	0	0	0	0
IND	Off_Bottom	1	0	1	1	1	0	0	1	1
IND	Off_Bottom	0	1	0	1	1	0	0	0	1
IND	Off_Bottom	1	0	1	0	1	0	0	0	0
IND	Off_Bottom	0	0	0	1	1	1	0	1	1
IND	Off_Bottom	0	1	0	0	1	0	0	1	1
IND	Off_Bottom	1	1	0	1	0	0	0	1	0
IND	Off_Bottom	0	0	0	0	0	0	0	1	0
IND	Off_Bottom	1	1	0	0	0	0	0	0	1
IND	Off_Bottom	0	0	1	0	1	1	0	0	0
SP	Off_Bottom	1	1	1	1	1	0	0	0	0
SP	Off_Bottom	1	1	1	1	0	0	0	0	1
SP	Off_Bottom	1	0	1	0	1	0	0	1	1
SP	Off_Bottom	1	1	1	1	0	0	0	0	1
SP	Off_Bottom	0	1	0	1	1	0	0	0	1
SP	Off_Bottom	1	1	0	1	0	0	0	1	0
SP	Off_Bottom	1	1	0	0	0	0	0	1	0
SP	Off_Bottom	1	1	0	0	0	0	0	1	1
SP	Off_Bottom	1	1	0	1	0	0	0	0	0
SP	Off_Bottom	0	1	0	1	1	0	0	1	0
VA	Off_Bottom	0	1	0	1	0	0	0	1	1
VA	Off_Bottom	0	0	1	0	1	0	0	0	1
VA	Off_Bottom	0	0	0	1	1	0	0	1	0
VA	Off_Bottom	0	1	0	1	0	0	0	1	0
VA	Off_Bottom	1	1	0	1	0	0	0	0	1
VA	Off_Bottom	1	1	0	0	0	0	0	0	0
VA	Off_Bottom	0	0	1	0	0	0	0	1	1
VA	Off_Bottom	0	0	1	0	0	0	0	0	1
VA	Off_Bottom	1	1	0	0	0	0	0	0	0
VA	Off_Bottom	1	1	1	1	1	0	0	1	1
VA	Off_Bottom	1	1	0	1	1	0	0	1	1
SP	Deep_Water	0	0	0	0	0	1	0	0	0

(continued)

Group	Method	Equipment	Disease	Seagrass	Climate	Winds	Grazers	Training	SpatialConflict	Health
SP	Deep_Water	0	0	0	0	1	0	1	0	0
SP	Deep_Water	0	0	0	0	0	0	1	0	0
SP	Deep_Water	0	0	0	0	0	1	0	0	0
SP	Deep_Water	0	0	0	0	1	1	1	0	1
SP	Deep_Water	1	0	0	0	0	1	1	1	0
SP	Deep_Water	0	1	0	0	1	0	1	0	0
SP	Deep_Water	0	0	0	0	1	0	0	1	1
SP	Deep_Water	1	1	0	0	0	1	0	0	0
SP	Deep_Water	0	0	0	0	0	0	1	0	0

S7: Percentage of Participants Spending Their Income from Seaweed Farming on Various Items

Item	% Participants
School fees	69.4%
Food	55.5%
Everyday household expenses	38.9%
Clothes	36.1%
Savings	30.5%
Personal use	27.8%
House renovations	25.0%
Extra household commodities	19.4%

References

Adger WN, Arnell NW, Tompkins EL (2005) Successful adaptation to climate change across scales. Glob Environ Change 15(2):77–86

Allison EH, Ellis F (2001) The livelihoods approach and management of small-scale fisheries. Mar Policy 25(5):377–388

Arnold S (2008) Seaweed, power, and markets: a political ecology of the Caluya Islands, Philippines (Major paper). York University, Toronto

Brown K, Westaway E (2011) Agency, capacity, and resilience to environmental change: lessons from human development, well-being, and disasters. Annu Rev Environ Resour 36:321–342

Brugere C, Msuya FE, Jiddawi N, Nyonje B, Maly R (2019) The introduction of an improved seaweed farming technology for women's empowerment, livelihoods and environmental protection. Institute of Marine Sciences, Zanzibar. https://doi.org/10.13140/RG.2.2.34671.69280

Brugere C, Msuya FE, Jiddawi N, Nyonje B, Maly R (2020) Can innovation empower? Reflections on introducing tubular nets to women seaweed farmers in Zanzibar. Gender, Technology and Development, pp 1–21

Bryceson I (2002) Coastal aquaculture developments in Tanzania. Sustainable and non-sustainable experiences. West Indian Ocean J Mar Sci 1:1–10

Cohen PJ, Lawless S, Dyer M, Morgan M, Saeni E, Teioli H, Kantor P (2016) Understanding adaptive capacity and capacity to innovate in social–ecological systems: applying a gender lens. Ambio 45(3):309–321

Crawford B (2002) Seaweed farming: an alternative livelihood for small-scale fishers. Coastal Resources Center, Narragansett

Critchley AT, Largo D, Wee W, Bleicher L'honneur G, Hurtado AQ, Schubert J (2004) A preliminary summary on Kappaphycus farming and the impact of epiphytes. Japanese J Phycol 52(Supplement):231–232

Davis E (2011) Seaweed farmer education: is it enough to sustain the industry? Analyzing the status of stakeholder investment in Muungoni and Jambiani, Unguja. Independent Study Project (ISP) Collection. 1195. https://digitalcollections.sit.edu/isp_collection/1195

Department of Fisheries Development (2019) Export of Seaweed for the period 2008–2018. Ministry of Agriculture, Natural Resources, Livestock and Fisheries

Duarte CM, Wu J, Xiao X, Bruhn A, Krause-Jensen D (2017) Can seaweed farming play a role in climate change mitigation and adaptation? Front Mar Sci 4:100

Eklöf JS, de la Torre Castro M, Adelsköld L, Jiddawi NS, Kautsky N (2005) Differences in macrofaunal and seagrass assemblages in seagrass beds with and without seaweed farms. Estuar Coast Shelf Sci 63(3):385–396

Eklöf JS, Msuya FE, Lyimo TJ, Buriyo AS (2012) Seaweed farming in Chwaka Bay: a sustainable alternative in aquaculture. People, nature and research in Chwaka Bay. WIOMSA, Zanzibar, pp 213–233

Ferrol-Schulte D et al (2015) Coastal livelihood vulnerability to marine resource degradation: a review of the Indonesian national coastal and marine policy framework. Mar Policy. Pergamon 52:163–171

Fröcklin S et al (2012) Seaweed mariculture as a development project in Zanzibar, East Africa: a price too high to pay? Aquaculture Elsevier 356–357:30–39

Fröcklin S, de la Torre-Castro M, Håkansson E, Carlsson A, Magnusson M, Jiddawi NS (2014) Towards improved management of tropical invertebrate fisheries: including time series and gender. PLoS One 9(3):e91161

Hassan IH, Othman WJ (2019) Seaweed (Mwani) farming as an adaptation strategy to impacts of climate change and variability in Zanzibar. In: Climate change and coastal resources in Tanzania. Springer, Cham, pp 53–68

IPPmedia (2019) Zanzibar in final stages to construct a giant seaweed processing plant. https://ippmedia.com/en/business/zanzibar-final-stages-construct-giant-seaweed-processing-plant. Accessed 8 Sept 2019

Jabeen H, Johnson C, Allen A (2010) Built-in resilience: learning from grassroots coping strategies for climate variability. Environ Urban 22(2):415–431

Largo DB, Fukami K, Nishijima T (1995) Occasional pathogenic bacteria promoting ice-ice disease in the carrageenan-producing red algae Kappaphycus alvarezii and Eucheuma denticulatum (Solieriaceae, Gigartinales, Rhodophyta). J Appl Phycol 7(6):545–554

Mahongo SB, Francis J (2010) Monthly variations in sea level at the island of Zanzibar. Western Indian Ocean J Mar Sci 9(1):1–16

Makame MO, Shackleton S (2019) Perceptions of climate variability and change in relation to observed data among two east coast communities in Zanzibar, East Africa. Clim Dev:1–13

Msuya FE (2006a) The impact of seaweed farming on the social and economic structure of seaweed farming communities in Zanzibar, Tanzania. World Seaweed Resources, Version, 1, p 27

Msuya FE (2006b, September) The seaweed cluster initiative in Zanzibar, Tanzania. In: Proceedings of the 3rd regional conference on innovation systems and innovative clusters in Africa, Dar es Salaam, Tanzania, pp 246–260

Msuya FE (2011) The impact of seaweed farming on the socioeconomic status of coastal communities in Zanzibar, Tanzania. World Aquacult 42(3):45–48

Msuya FE, Porter M (2014) Impact of environmental changes on farmed seaweed and farmers: the case of Songo Songo Island, Tanzania. J Appl Phycol 26(5):2135–2141

Msuya FE, Shalli MS, Sullivan K, Crawford B, Tobey J, Mmochi AJ (2007) A comparative economic analysis of two seaweed farming methods in Tanzania. The Sustainable Coastal Communities and Ecosystems Program. Coastal Resources Center, University of Rhode Island and the Western Indian Ocean Marine Science Association.

Msuya FE, Buriyo A, Omar I, Pascal B, Narrain K, Ravina JJ, Mrabu E, Wakibia JG (2014) Cultivation and utilisation of red seaweeds in the Western Indian Ocean (WIO) region. J Appl Phycol 26(2):699–705

Namudu MT, Pickering TD (2006) Rapid survey technique using socio-economic indicators to assess the suitability of Pacific Island rural communities for Kappaphycus seaweed farming development. In: Eighteenth international seaweed symposium. Springer, Dordrecht, pp 15–23

Newman RJS, Capitani C, Courtney-Mustaphi C, Thorn JPR, Kariuki R, Enns C, Marchant R (2020) Integrating insights from social-ecological interactions into sustainable land use change scenarios for Small Islands in the Western Indian Ocean. Sustainability 12(4):1340

Osman-Elasha B, Goutbi N, Spanger-Siegfried E, Dougherty B, Hanafi A, Zakieldeen S, Sanjak A, Atti HA, Elhassan HM (2006) Adaptation strategies to increase human resilience against climate variability and change: lessons from the arid regions of Sudan. Assessments of impacts and adaptations to climate change (AIACC) working paper, 42

Quandt A (2018) Measuring livelihood resilience: the household livelihood resilience approach (HLRA). World Dev 107:253–263

Revolutionary Government of Zanzibar (2013) Zanzibar climate change strategy. http://www.paulwatkiss.co.uk/newimagesanddocs/ZanzibarSummaryLRdraftfinal.pdf. Accessed 2 Sept 2019

Salafsky N, Wollenberg E (2000) Linking livelihoods and conservation: a conceptual framework and scale for assessing the integration of human needs and biodiversity. World Dev 28(8):1421–1438

Sandelowski M (1995) Sample size in qualitative research. Res Nurs Health 18(2):179–183

Scoones I (1998) Sustainable rural livelihoods: a framework for analysis. IDS working paper no. 72, Institute of Development Studies, Brighton

Serrat O (2017) The sustainable livelihoods approach. In: Knowledge solutions. Springer Singapore, Singapore, pp 21–26

Sietz D, Boschutzz M, Klein RJT (2011) Mainstreaming climate adaptation into development assistance: rationale, institutional barriers and opportunities in Mozambique. Environ Sci Policy 14(4):493–502

Sievanen L, Crawford B, Pollnac R, Lowe C (2005) Weeding through assumptions of livelihood approaches in ICM: seaweed farming in the Philippines and Indonesia. Ocean Coast Manag 48 (3–6):297–313

Spencer L, Ritchie J, O'Connor W (2003) Analysis: practices, principles and processes. Qualitative research practice: a guide for social science students and researchers, SAGE, London, 199, p 218

Suckall N, Tompkins E, Stringer L (2014) Identifying trade-offs between adaptation, mitigation and development in community responses to climate and socio-economic stresses: Evidence from Zanzibar, Tanzania. Appl Geogr Pergamon 46:111–121

Temple B, Young A (2004) Qualitative research and translation dilemmas. Qual Res 4(2):161–178

Tobisson E (2014) Consequences and challenges of tourism and seaweed farming: a narrative on a coastal community in Zanzibar. Western Indian Ocean J Mar Sci 12(2):169–184

Tsiresy G, Preux J, Lavitra T, Dubois P, Lepoint G, Eeckhaut I (2016) Phenology of farmed seaweed Kappaphycus alvarezii infestation by the parasitic epiphyte Polysiphonia sp. in Madagascar. J Appl Phycol 28(5):2903–2914

Vairappan CS (2006) Seasonal occurrences of epiphytic algae on the commercially cultivated red alga Kappaphycus alvarezii. J Appl Phycol 18(3–5):611–617

Valderrama D et al (2015) The economics of Kappaphycus seaweed cultivation in developing countries: a comparative analysis of farming systems. Aquac Econ Manag 19(2):251–277

Wallevik HB, Jiddawi N (2001) Impacts of tourism on the activities of the women of the southeast coast of Unguja, Zanzibar. In Marine Science Development in Tanzania and Eastern Africa. Proceedings of the 20 th Anniversary Conference on Advances in Marine Science in Tanzania 28:535–550

Wodon Q, Blackden CM eds., (2006) Gender, time use, and poverty in sub-Saharan Africa. The World Bank

Wright H, Vermeulen S, Laganda G, Olupot M, Ampaire E, Jat ML (2014) Farmers, food and climate change: ensuring community-based adaptation is mainstreamed into agricultural programmes. Clim Dev 6(4):318–328

ZaSCi (2019) Zanzibar Seaweed cluster initiative. https://www.mzfn.org/Advocacy_For_Seaweed.html. Date accessed 2/09/19

7

Determinants of Cattle Farmers' Perception of Climate Change in the Dry and Subhumid Tropical Zones of Benin (West Africa)

Yaya Idrissou, Alassan Seidou Assani, Mohamed Nasser Baco and Ibrahim Alkoiret Traoré

Contents

Abstract

Understanding the factors influencing the perception of climate change can help improve policies for strengthening the adaptive capacity of pastoralists with regard to climate change. Despite this importance, few studies have focused on this issue,

---

Y. Idrissou (✉) · A. S. Assani · I. Alkoiret Traoré
Laboratoire d'Ecologie, Santé et Production Animales (LESPA), Faculté d'Agronomie, Université de Parakou, Parakou, République du Bénin
e-mail: yayaidriss2617@gmail.com; alassanassani@gmail.com; alkouarit@gmail.com

M. N. Baco
Laboratoire Société-Environnement (LaSEn), Faculté d'Agronomie, Université de Parakou, Parakou, République du Bénin
e-mail: nasserbaco@yahoo.fr

especially among cattle farmers. In order to attempt filling this gap, this study analyzed the determinants of the perception of climate change by cattle farmers distributed in the dry and sub-humid tropical zones of Benin as well as the current adaptation strategies developed by these farmers. For this purpose, surveys were carried out through group discussions and an individual questionnaire administered to 360 cattle farmers in the two climatic zones. The data collected related to the sociodemographic characteristics of cattle farmers and their perception of climate change and adaptation strategies. A binary logit model has identified the factors that influence cattle farmers' perceptions of climate change. The results of the study showed that cattle farmers perceive a drop in rain (at least 77%), an increase in temperature (at least 80%), and violent winds (at least 60%). Breeding experience, level of education of the farmer, household size, membership of a breeders' organization, and cattle herd size determine these perceptions. Four major groups of adaptation strategies have been developed by farmers to cope with climate change. These are production adjustment strategies, activity diversification strategies, livestock management strategies, and selection strategies. The political implication of this study is that government and development partners should integrate these factors into projects and programs related to climate change.

Keywords

Animal · Climatic zone · Climate change · Adaptation strategies · West Africa

Introduction

Climate change is currently an increasingly visible threat to the viability of the rural population of sub-Saharan Africa, where communities depend mainly on the exploitation of natural resources (Kaboré et al. 2019; Adimassu and Kessler 2016). Livestock rearing is one of the main economic activities on which the poorest people in sub-Saharan Africa depend as a source of food and income. Despite its importance, livestock is currently threatened by climate change (Apata et al. 2009; Deressa et al. 2009) because of its high dependence on natural resources specifically fodder and water (Idrissou et al. 2019; IUCN 2010). The impacts of climate change in the livestock sector are felt in the production and quality of forage crops (Polley et al. 2013; Chapman et al. 2012), water availability, animal growth, milk production, reproduction, and disease (Henry et al. 2012). Faced with this situation, the challenge for the scientific community is to produce knowledge enabling farmers to anticipate the effects of climate change on their system and to develop methods and tools to adapt to it (Sautier 2013). To achieve this, it is necessary to understand how pastoralists perceive climate change (Deressa et al. 2011), as this influences the way they manage climate-related risks and opportunities, as well as the strategies put in place to adapt (Mamba 2016). In developing countries, numerous studies have dealt with the perception and adaptation strategies of pastoralists in the face of climate change (Idrissou et al. 2020; Sanou et al. 2018; Ayanlade et al. 2017). In Benin, for example, several studies show that pastoralists perceive climate change through the drop in rain, irregular rainfall, and late start and early end of the rainy season (Idrissou et al. 2020; Dossa et al. 2017; Zakari et al. 2015). Pastoralists in Burkina Faso have unanimously discerned some changes in precipitation and temperature (Sanfo et al. 2015). They saw a decrease in annual precipitation, an increase in the intensity of precipitation, and the frequency of flooding. In Kenya, pastoralists have reported changes in the amount and distribution of precipitation, fog, temperature, and wind over the past 20 to 30 years (Cuni-Sanchez et al. 2019). To cope with the harmful effects of climate change, pastoralists in developing countries have developed several strategies such as herd mobility, storage of crop residues, and integration of livestock rearing with crop farming, among others (Idrissou et al. 2019).

These studies, although they allow apprehending the perceptions and adaptation strategies of cattle farmers, are still insufficient. Indeed, information on the factors that determine the perceptions of pastoralists has not often been analyzed. Knowledge of this aspect is important for science and will allow better targeting of policies to support the adaptation of pastoralists to climate change in developing countries.

Benin, a small country in West Africa, has three climatic zones, the most vulnerable to climate change being the dry and sub-humid tropical zones (Gnanglè et al. 2011; MEHU 2011). It is in these most vulnerable areas that more than 85% of the country's cattle farms are concentrated (Alkoiret et al. 2011). These cattle farms will be severely affected by the effects of climate change, resulting in reduced productivity (Nardone et al. 2010). Pastoralists in these zones are thus exposed to risks of food insecurity and increasing poverty. A study analyzing the perception and adaptation strategies of pastoralists in these areas in the face of climate change is timely as it can help improve policies aimed at supporting these pastoralists to adapt sustainably to climate change (Folefack and Tenikue 2015; Mabe et al. 2014). The aims of this study are therefore to (i) analyze the determinants of the perception of climate change by cattle farmers in the dry and sub-humid tropical zones of Benin and (ii) identify the current adaptation strategies developed by these farmers.

Material and Methods

Study Areas

This study was carried out in two of the three climatic zones of Benin (located 6° and 12° 50′N and 1° and 3° 40′ E). These are the dry tropical zone between 9° 45′ and 12° 25′ N and the sub-humid tropical zone located between 7° 30′ and 9° 45′ N (Fig. 1). The choice of these zones is based on the fact that climate forecasts indicate that they are the most vulnerable to rainfall deficit and high sunshine (Gnanglè et al. 2011; MEHU 2011), yet more than 85% of the Beninese cattle herd is concentrated in these zones (Alkoiret et al. 2011).

In each zone, two (2) municipalities were chosen based on the large number of cattle farmers and preliminary interviews with technicians from the "Agences Territoriales pour le Développement Agricole" (ATDA). Thus, the municipalities

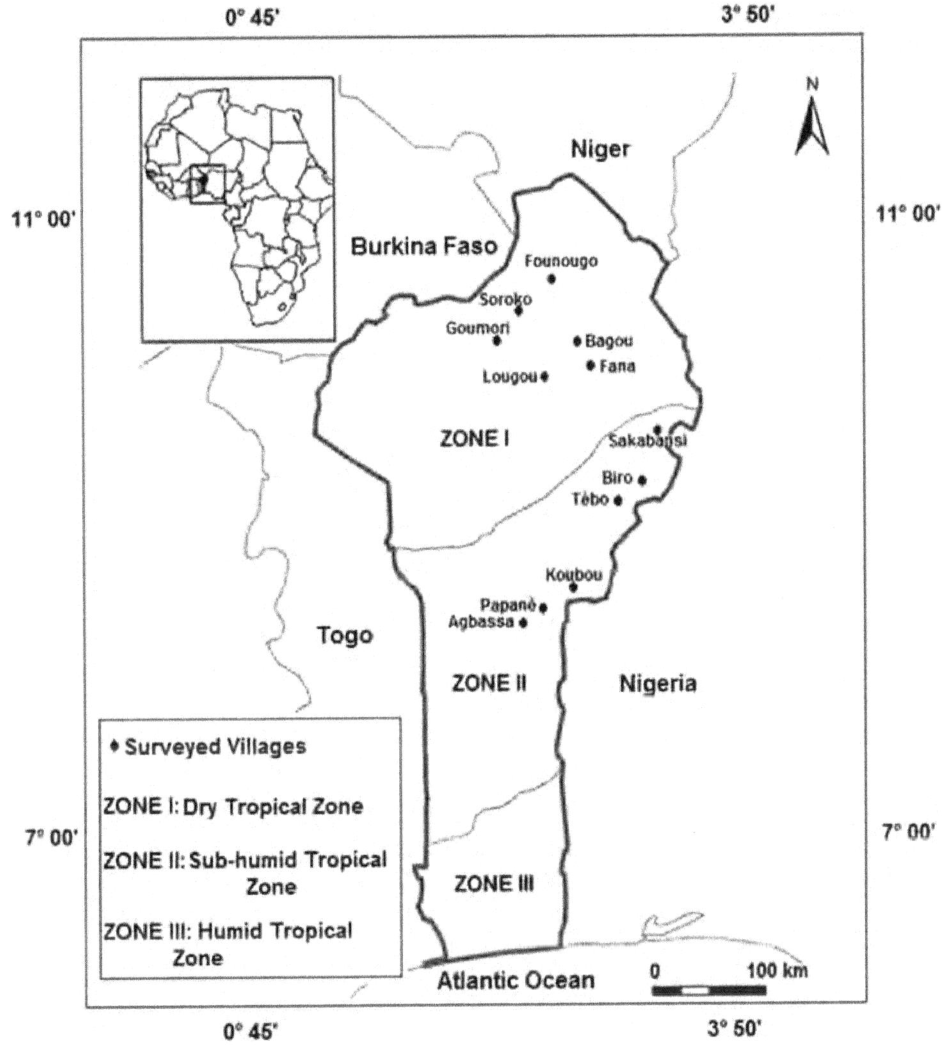

Fig. 1 Location of the villages surveyed in Benin

of Tchaourou and Nikki were selected in the sub-humid tropical zone and those of Gogounou and Banikoara in the dry tropical zone. Within each municipality, three villages have been selected (Fig. 1). In the dry tropical zone, which includes the municipalities of Gogounou and Banikoara, the mean annual rainfall is often less than 1000 mm, and the relative humidity varies from 18 to 99% (highest in August). The temperature varies from 24 °C to 31 °C. The soils in this zone are hydromorphic, well-drained soils, and lithosols. The vegetation of this zone is mainly composed of savannas with trees of smaller size. The sub-humid tropical zone, made up of the municipalities of Tchaourou and Nikki, is characterized by unimodal precipitation, from May to October, and lasts about 113 days with an annual mean rainfall varying between 900 and 1110 mm. The annual temperature ranges from 25 °C to 29 °C and the relative humidity from 31% to 98%. Soils in this zone are ferruginous with variable fertility. The vegetation of the sub-humid tropical zone is characterized by a mosaic of woodland, dry dense forests, tree and shrub savannas, and gallery forests.

Data Collection

Data collection was carried out from November 2018 to April 2019 in two stages that include the exploratory study and in-depth interviews.

Exploratory Study

During this phase, interviews were carried out with technicians from the ATDA in order to identify the villages and cattle farmers to be surveyed. On the basis of the defined criteria, with the local technicians, the villages of Koubou, Papanè, and Agbassa were identified in the municipality of Tchaourou; the villages Tèbo, Biro, and Sakabansi in the municipality of Nikki; the villages of Bagou, Fana, and Lougou in the municipality of Gogounou; and finally those of Founougo, Goumori, and Soroko in the municipality of Banikoara (Fig. 1).

After identifying the villages, focus groups of staff varying from 6 to 15 people were carried out (one focus group per village). During the focus groups, the questions were open and made it possible to record the maximum of responses on the perceptions and adaptation strategies of cattle farmers. At the end of these focus groups, a list of climatic parameters cited by cattle farmers was drawn up. These climatic parameters have been broken down into different indicators of their manifestations as cited by farmers. A global synthesis was made to constitute the content of the questionnaire for in-depth interviews.

The interviews carried out during the exploratory study made it possible to randomly select 30 cattle farmers per village to whom questions were addressed individually for the continuation of the study. Thus, a total of 360 cattle farmers were surveyed during this study. The criteria for choosing cattle farmers were having cattle breeding as their main activity and being at least 50 years old. The choice of breeding as the main activity to discriminate the respondents is justified by the fact that several studies focus on agro-pastoralists and generalize the results obtained both to agro-pastoralists and to pastoralists. However, these socio-professional categories face different socioeconomic problems (Zampaligré et al. 2014). The age barrier (50 years) is explained by the fact that climate change is very slow and elderly people are needed to have reliable historical information (Kaboré et al. 2019; Bambara et al. 2016).

In-Depth Interviews

The in-depth interview consisted of collecting data through semi-structured interviews with the 360 cattle farmers identified during the exploratory study. For data collection, local investigators were recruited and trained. The training of local investigators was undertaken for a week and piloted before the start of interviews with cattle farmers. The aim of the training was to minimize bias and errors in data collection. The interviewers used to conduct the interviews with the cattle farmers were selected from each study village and understood the local language of the village. The data were collected using a questionnaire. A first series of questions related to the sociodemographic characteristics of the cattle farmers (sex, age, breeding experience, ethnic group, household size, number of agricultural assets, level of education, contact with agricultural extension services, membership of a breeders' organization, etc.) and a second series of questions concerned the perception of climate change indicators as well as the adaptation strategies developed. The questions posed to farmers on their perception of climate change are consistent with the indices of the Expert Team on Climate Change Detection Monitoring and Indices (ETCCDMI) (Zhang and Yang 2004). Climate change indicators are meteorological parameters whose evolution over time reflects climate change. These indicators are total annual precipitation, rainfall intensity, daily maximum, and minimum temperatures. Other parameters were also taken into account including "rainless days" or pockets of drought, vortex, and strong winds (Bambara et al. 2016; Salack et al. 2012).

Statistical Analysis

The data from the surveys were processed using the Statistical Package for Social Sciences (SPSS) version 17. Frequencies of responses were reported and compared with chi-square ($\chi 2$) test. The quantitative variables describing the cattle farmers surveyed were presented as means ± standard deviations and compared between climatic zones using the Mann-Whitney U nonparametric test (McDonald 2009).

The determinants that influence cattle farmers' perception of climate change have been analyzed by binary logistic regression (Kaboré et al. 2019; Uddin et al. 2017). The equation of the binary model is as follows:

$$Yi = Xi\,\beta + \varepsilon i \quad (1)$$

where Yi is the variable which takes the value 1 if the farmer perceives a climate change indicator and 0 if he does not perceive it; Xi is the set of explanatory variables indicating the factors that influence the cattle farmers' perception of climate change; and εi is the standard error.

Before estimating the logistic regression model, the explanatory variables were checked to determine the existence of multi-colinearity, using the contingency coefficient test (Uddin et al. 2017). A collinearity was observed between the breeding experience and age; between the number of agricultural assets and household size; and between membership of a breeders' organization and contact with extension services. Consequently, age, number of agricultural assets, and contact with extension services were omitted from the logistic regression model after the multi-colinearity test. The explanatory variables used for the regressions are sex, breeding experience, ethnic group, level of education, location (climatic zone), membership to a breeders' organization, household size, and herd size.

Results and Discussion

Sociodemographic Characteristics of the Cattle Farmers Surveyed

The sociodemographic characteristics of the cattle farmers surveyed are summarized in Table 1. The majority of the cattle farmers surveyed are of the Fulani sociocultural group (80.83%) and of the male sex (92.5%). They are relatively old (56.43 ± 5.88 years) and have an average experience of 30 years in cattle breeding. Cattle farmers in the sub-humid tropical zone were older and more experienced ($p < 0.05$) than those in the dry tropical zone. Very few cattle farmers have been educated (5%). The household size of the cattle farmers surveyed was on average 11 people, and the cattle herd size was on average 64 heads. A large number of cattle farmers are members of an organization (90.92%) and are also in contact with agricultural extension services (66.21%). It is specified that the number of cattle farmers in contact with agricultural extension services in the dry tropical zone is significantly high ($p < 0.05$) compared to that of the sub-humid tropical zone. This could be explained by the fact that historically the dry zone is an area purely dedicated to animal husbandry, which leads technicians leaving agricultural college and university to settle more in this zone. Additionally, the first livestock extension structures were created in the dry zone.

Table 1 Sociodemographic characteristics of cattle farmers

Variables	Climatic zones		Total
	STZ	DTZ	
Percentage (%)			
Sex			
Male	95.00^{a}	90.00^{a}	92.50
Female	5.00^{a}	10.00^{a}	7.50
Ethnic group			
Fulani	80.00^{a}	81.67^{a}	80.83
Bariba	20.00^{a}	18.33^{a}	19.17
Level of education			
Educated	6.67^{a}	3.33^{a}	5.00
Non-educated	93.33^{a}	96.67^{a}	95.00
Membership in an organization	88.33^{a}	93.50^{a}	90.92
Contact with the extension	60.67^{a}	71.75^{b}	66.21
Mean ± standard deviation			
Age	57.95 ± 6.52^{a}	54.91 ± 4.7^{b}	56.43 ± 5.88
Breeding experience	34.37 ± 10.95^{a}	25.75 ± 8.4^{b}	30.05 ± 10.63
Number of agricultural assets	8.55 ± 3.64^{a}	7.16 ± 3.25^{a}	7.85 ± 3.50
Household size	11.70 ± 4.64^{a}	11.71 ± 4.81^{a}	11.70 ± 4.71
Cattle herd size	61.28 ± 41.59^{a}	67.72 ± 35.46^{a}	64.50 ± 38.62

STZ sub-humid tropical zone, *DTZ* dry tropical zone

[a, b]The values of the same line indexed by different letters are significantly different at the 5% threshold

Cattle Farmers' Perception of Climate Change Indicators

Seventy-seven percent (77%) of cattle farmers in the sub-humid tropical zone and 82% in the dry tropical zone perceived a decrease in rainfall (Fig. 2a). The indicator "high intensity of rainfall" was perceived by 78% of the cattle farmers of the dry tropical zone against 52% of the cattle farmers of the sub-humid tropical zone. In the both zones, indicators of change in precipitation such as the duration of the drought pocket, the late onset, and the early cessation of the rains were perceived by more than 80% of the farmers interviewed. At least 50% of the farmers in both zones perceived an irregularity in the rains. These indicators of change in precipitation perceived by pastoralists could be explained by the fact that these indicators remain the most visible in the observation of rainfall pejoration in Africa (Bambara et al.

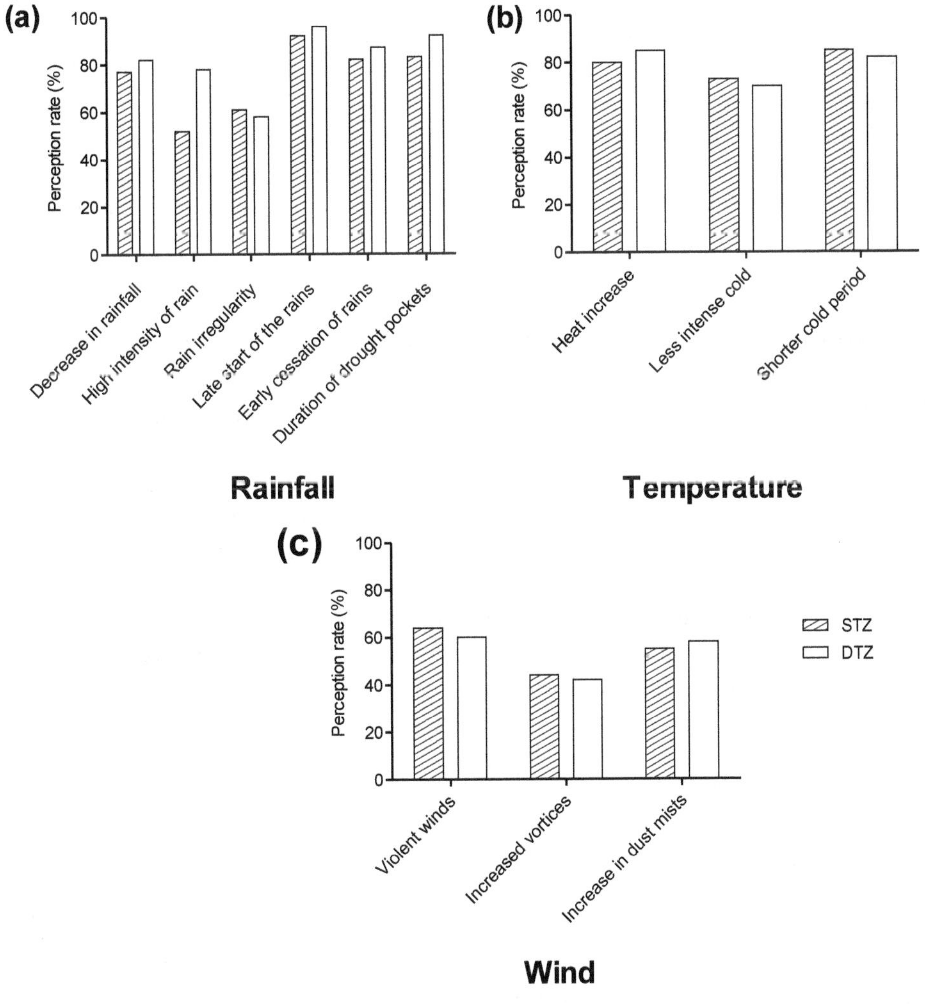

Fig. 2 Perception of indicators of changes in rainfall (**a**) in temperature (**b**) and wind (**c**) by cattle farmers in the dry (white) and sub-humid (black) tropical zones of Benin

2013). These results therefore confirm the work of several authors in Africa (Cuni-Sanchez et al. 2019; Opiyo et al. 2016).

The majority (at least 80%) of cattle farmers in the both zones recognize that the heat is getting stronger (Fig. 2b). These results reflect an increase in maximum and minimum daily temperatures observed throughout the year. Farmers also indicated that the cold season is warming up (70% and 73% of the cattle farmers in the dry and sub-humid tropical zones respectively) and tends to become shorter (82% in the dry zone and 85% in the sub-humid zone). As the temperature is a direct feeling, the farmers can easily see its increase through warmer days and nights. In addition, high temperatures cause animals to seek more shade and water (Idrissou et al. 2020). This behavior of animals observed by farmers could also explain their perception of the increase in temperature.

According to 60% of cattle farmers in the dry tropical zone and 64% of the farmers in the sub-humid tropical zone, the winds have become more and more violent (Fig. 2c). The increase in vortices was reported by at least 40% of cattle farmers in the both zones. In addition, more than 50% of the cattle farmers in the two zones perceived an increase in dust mists. Strong winds cause material damage such as destruction of roofs, erosion of cultivable land, uprooting of trees, etc. This damage noted by farmers could explain their perception. Similar perceptions have been reported by several authors in other parts of Africa (Limantol et al. 2016; Opiyo et al. 2016).

Determinants of Cattle Farmers' Perception of Climate Change

The analysis of binary logistic regression performed to determine the factors that influence the farmers' perception of the change in precipitation and temperature is summarized in Table 2. The variable "Breeding Experience" positively affects perceptions of reduced rainfall and late onset and early cessation of rains. This result indicates that farmers with longer years of cattle breeding experience were more likely to perceive climate change. In addition, experienced farmers observe changes over time and compare them to current climatic conditions, allowing them to quickly perceive climate change. This result is similar to those obtained by several authors (Sanogo et al. 2017; Uddin et al. 2017).

The level of education of the cattle farmers positively influences the perception of the late onset of the rains at the threshold of 10% (Table 2). The most educated farmers are more interested in calendar dates or the start of school holidays, which generally coincide with the start of winter (Kaboré et al. 2019). The less educated do not really make the difference between an early or late season.

Membership of a breeders' organization influences cattle farmers' perception of reduced rains and late onset and early cessation of rains. Breeders' organizations benefit from training from development partners through non-governmental organizations (NGOs), agricultural development projects, and programs. Through these different structures, cattle farmers are made aware of climate change as well as the present and future consequences on their livelihood. In addition to these sources of

Table 2 Determinants of cattle farmers' perception of change in precipitation and temperature

Variables	Decrease in rainfall			Late start of the rains			Early cessation of rains			Heat increase		
	β	$SE\ \beta$	p	β	$SE\ \beta$	p	β	$SE\ \beta$	p	β	$SE\ \beta$	p
Constant	0.224*	2.458	0.0927	7.020***	2.528	0.005	3.887*	2.329	0.095	4.956	3.346	0.139
Climatic zone	0.652	0.522	0.212	0.202	0.459	0.659	0.277	0.459	0.546	1.051	0.608	0.840
Breeding experience	0.077*	0.043	0.074	0.093**	0.039	0.018	0.077**	0.037	0.041	0.017	0.056	0.763
Sex	2.210	0.992	0.26	1.223	0.800	0.126	0.021	0.735	0.977	1.543	1.204	0.20
Ethnic group	0.020	0.543	0.97	0.782	0.502	0.12	−0.524	0.504	0.298	0.291	0.636	0.647
Education level	1.536	1.224	0.209	0.363*	0.930	0.0696	0.925	0.954	0.332	0.893	1.187	0.452
Household size	0.029	0.046	0.521	0.018	0.043	0.673	0.023	0.042	0.579	0.117**	0.056	0.039
Membership in an organization	0.351*	0.614	0.0568	−0.246*	0.571	0.0666	−0.426**	0.561	0.0447	0.091	0.751	0.903
Cattle herd size	0.004**	0.005	0.0382	0.061**	0.050	0.042	0.008*	0.005	0.0706	0.011*	0.006	0.087
Number of observations	360			360			360			360		
−2log likelihood	14.33			153.56			154.94			102.56		
Prob > Chi	0.009			0.019			0.025			0.000		

*significant value at 10% ($0.05 < p \leq 0.10$); **significant value at 5% ($0.01 < p \leq 0.05$); ***significant value at 1% ($p < 0.01$)

learning are the relationships that cattle farmers have with each other which serve as channels for sharing experiences.

Household size positively influenced the cattle farmers' perception of the increased heat. This finding indicates that with increasing household size, the likelihood that cattle farmers perceive climate change increases. Milk represents an essential constituent of the food ration of Fulani cattle farmers and also generates income for their households (Alkoiret et al. 2010). Climate change is causing a decrease in the milk production of animals (Henry et al. 2012). Large households are therefore no longer able to meet the milk demand of their large families. This is why cattle farmers can more easily detect climate change (Kosmowski et al. 2016).

Cattle herd size has also influenced cattle farmers' perception of the increase in heat, the decrease in rainfall, and the late onset and the early cessation of rains. This means that farmers with large numbers of cattle perceive climate change better than those with small numbers due to the high demand for water and forage. This result is similar to that obtained among Turkana cattle farmers in northwestern Kenya (Opiyo et al. 2016).

Cattle Farmers' Adaptation Strategies to Climate Change

To adapt to climate change, cattle farmers in the dry and sub-humid tropical zones of Benin have developed several strategies (Table 3). These strategies can be grouped

Table 3 Adaptation strategies of cattle farmers in the dry and sub-humid tropical zones of Benin in the face of climate change

Adaptation strategies			Climatic zones			
Strategy group	Type of strategy	Total	DTZ	STZ	X^2	p-Value
Production adjustment strategies	Forage cropping	31.1	10^{a}	52.2^{b}	72.9	<0.0001
	Storage of crop residues	76.1	82.2^{a}	70^{b}	6.7	0.009
	Making hay	18.9	19.4^{a}	18.3^{a}	0.01	0.89
	Night grazing	14.4	13.9^{a}	15^{a}	0.02	0.88
	Use concentrated feed	72.8	77.8^{a}	67.8^{b}	4.05	0.04
Activities diversification strategies	Integration livestock-crop	83.9	95^{a}	72.8^{b}	31.2	<0.0001
	Fattening	15.3	11.1^{a}	19.4^{a}	3.68	0.051
	Off-farm activities	25.6	15^{a}	36.1^{b}	19.9	<0.0001
Livestock management strategies	Herds destocking	70	77.8^{a}	62.2^{b}	9.6	0.001
	Livestock diversification	58.9	92.8^{a}	25^{b}	167.9	<0.0001
	Pastoral mobility	96.1	100^{a}	92.2^{b}	12.5	0.0003
Selection strategies	Breeding local breeds	2.3	19.4^{a}	21.1^{a}	0.06	0.79
	Cross between local breed and breed adapted to heat	25.3	31.7^{a}	18.9^{b}	7.1	0.007

DTZ dry tropical zone, *STZ* sub-humid tropical zone

$^{a,\,b}$The values of the same line indexed by different letters are significantly different at the 5% threshold into four major groups of strategies, namely, production adjustment strategies, activities diversification strategies, livestock management strategies, and selection strategies. This classification is similar to that developed by Calvosa et al. (2010).

The production adjustment strategies consist of the storage of crop residues, use of concentrated feed, forage cropping, making hay, and night grazing. The adoption rates for the storage of crop residues (82.2%) and the use of feed concentrates (72.8%) were significantly higher ($p < 0.05$) in the dry tropical zone than in the sub-humid tropical zone. Conversely, forage cropping (52.2%) was more adopted ($p < 0.0001$) by cattle farmers in the sub-humid tropical zone. Indeed, the rainfall and the number of rainy days recorded in the sub-humid zone being higher than that obtained in the dry zone could favor the forage cropping in this zone. Also, as the climatic conditions are not conducive to the forage cropping in the dry zone, this leads cattle farmers in this zone to resort to other strategies such as the use of feed concentrates and the storage of crop residues to feed the animals.

Integration of livestock rearing and crop cultivation, fattening, and off-farm activities are the types of strategies contained in the activities under the diversification strategies category. The practice of fattening did not vary significantly ($p > 0.05$) from one climatic zone to another. On the other hand, the integration livestock and crop was more adopted ($p < 0.0001$) in the dry zone than in the sub-humid zone. This result means that pastoralists have moved to agro-pastoralism. Indeed, milk was the staple food for Fulani cattle farmers (Alkoiret et al. 2010), but with the fall in milk production due to climate change, the consumption of cereals has increased. To obtain these cereals, cattle farmers exchanged milk for these products with farmers. But today, given the decrease in crop yield and the strong demand due to the demographic surge, cereals are inaccessible because their prices have increased, making the exchange of products difficult. This may be the reason why cattle farmers integrate livestock and crop because they are convinced that livestock cannot be their only source of food. This result is similar to those obtained in Burkina Faso by Sanfo et al. (2015). Off-farm activities were more practiced by cattle farmers in the sub-humid tropical zone ($p < 0.0001$) than those in the dry zone. This could be explained by the fact that the sub-humid tropical zone is an area which abounds in urban centers of a commercial nature, which could facilitate trade. Thus, income from off-farm activities can be used by cattle farmers to increase the level of investment in inputs such as labor, feed concentrates, and veterinary products.

Herd destocking, livestock diversification, and pastoral mobility, all of which are livestock management strategies, were adopted more in the dry zone ($p \leq 0.001$) than in the sub-humid zone. Indeed, the insufficiency of forage due to the precariousness of the rains pushes the cattle farmers having a large number of cattle to practice the transhumance to reduce the risk of mortality (Kiema et al. 2013). In addition, other cattle farmers are forced to reduce the size of their herd (Kima et al. 2015; Oyekale 2014). Livestock diversification through the introduction of small ruminants into the breeding constitutes a real advantage for the farmers because of their low food needs, their larger feeding areas, and their higher reproduction rates (IUCN 2010).

Breeding local breeds adapted to local climatic conditions, and the crossing between local breeds and heat-resistant breeds were the two selection strategies implemented by the cattle farmers in both zones. The adoption rates for the first strategy were not significantly different ($p < 0.05$) in the two study zones. On the other hand, crossbreeding between local breeds and heat-resistant breeds as an adaptation strategy was more adopted in the dry zone ($p = 0.007$) than in the sub-humid zone. This could be explained by the fact that the dry zone is a pre-Sahelian zone where there is a high temperature, therefore requiring the breeding of heat-resistant animals.

Conclusion

The manifestations of climate change are perceived by cattle farmers in the dry and sub-humid tropical zones of Benin. The changes in rainfall are felt through signs such as reduced rainfall and late start and early end of the rainy seasons. Changes in temperature and wind were felt through increased heat and violent winds. This study showed that the sociodemographic characteristics of cattle farmers such as the level of education and membership of an organization influence local perceptions of climate change.

Adaptation strategies implemented by cattle farmers can be grouped into four main groups. These are production adjustment strategies, activities diversification strategies, livestock management strategies, and selection strategies. These current developed strategies allow farmers to take advantage of their livelihood.

Based on the results of this study, there is a need to strengthen the adaptive capacities of farmers in both zones through their access to education and training on adaptation to climate change within breeders' organizations. These help to improve their perception of this phenomenon and help them to better develop adaptation strategies. Climate information is needed to enable them to increase production to achieve food security. In addition, endogenous climate change indicators should be further promoted, as they allow farmers to predict the course of the rainy season and guide them better in implementing their adaptation strategies to climate change.

Acknowledgments Yaya Idrissou thanks the International Foundation for Science for its financial assistance (IFS, B/6189-1). He also thanks the Intergovernmental Panel on Climate Change (IPCC) and the Prince Albert II of Monaco Foundation (FPAM) for their financial support. The contents of this document are solely the liability of Yaya Idrissou and under no circumstances may be considered as a reflection of the position of the Prince Albert II of Monaco Foundation and/or the IPCC.

References

Adimassu Z, Kessler A (2016) Factors affecting farmers' coping and adaptation strategies to perceived trends of declining rainfall and crop productivity in the central rift valley of Ethiopia. Environ Syst Res 5:13

Alkoiret IT, Awohouedji DYG, Gbangboche AB, Bosma RH (2010) Productivité des systèmes d'élevage bovin de la commune de Gogounou au nord-est du Bénin. Ann Sci Agron 14:145–163

Alkoiret TI, Radji M, Babatoundé S (2011) Typologie des élevages bovins installés dans la commune de Ouaké au nord-ouest du Bénin. Livest Res Rural Dev 23. http://www.lrrd.org/lrrd23/3/alko23050.htm

Apata TG, Samuel KD, Adeola AO (2009) Analysis of climate change perception and adaptation among arable food crop farmers in South Western Nigeria. Contributed paper prepared for presentation at the International Association of Agricultural Economists'2009 Conference, Beijing, 16–22 Aug 2009

Ayanlade A, Radeny M, Morton JF (2017) Comparing smallholder farmers' perception of climate change with meteorological data: a case study from southwestern Nigeria. Weather Clim Extrem 15:24–33

Bambara D, Bilgo A, Hien E, Masse D, Thiombiano A, Hien V (2013) Perceptions paysannes des changements climatiques et leurs conséquences socio environnementales à Tougou et Donsin, climats sahélien et sahélo-soudanien du Burkina Faso. Bull Rech Agron Bénin 74:8–16

Bambara D, Thiombiano A, Hien V (2016) Changements climatiques en zones nord-soudanienne et sub-sahelienne du Burkina Faso: comparaison entre savoirs paysans et connaissances scientifiques. Rev Décologie 71:35–58

Calvosa C, Chuluunbaatar D, Fara K (2010) Livestock and climate change. Livestock Thematic Papers

Chapman SC, Chakraborty S, Dreccer MF, Howden SM (2012) Plant adaptation to climate change—opportunities and priorities in breeding. Crop Pasture Sci 63:51–268

Cuni-Sanchez A, Omeny P, Pfeifer M, Olaka L, Mamo MB, Marchant R, Burgess ND (2019) Climate change and pastoralists: perceptions and adaptation in montane Kenya. Clim Dev 11:513–524

Deressa TT, Hassan RM, Ringler C, Alemu T, Yesuf M (2009) Determinants of farmers' choice of adaptation methods to climate change in the Nile Basin of Ethiopia. Glob Environ Chang 19:248–255

Deressa TT, Hassan RM, Ringler C (2011) Perception of and adaptation to climate change by farmers in the Nile basin of Ethiopia. J Agric Sci 149:23–31

Dossa LH, Lesse DP, Houinato M, Sinsin B, Souberou F, Yabi I (2017) Vulnérabilité de l'élevage transhumant à la Variabilité hydro-climatique dans le nord-est de la République du Bénin. Rev CAMES 8:233–249

Folefack DP, Tenikue M (2015) Choice of coping strategies of producers facing the cotton crisis: a multinomial logit analysis in Cameroon. Int J Innov Appl Stud 11:77–89

Gnanglè CP, Glèlè Kakaï R, Assogbadjo AE, Vodounnon S, Afouda Yabi J, Sokpon N (2011) Tendances climatiques passées, modélisation, perceptions et adaptations locales au Bénin. Climatologie 8:27–40

Henry B, Charmley E, Eckard R, Gaughan JB, Hegarty R (2012) Livestock production in a changing climate: adaptation and mitigation research in Australia. Crop Pasture Sci 63:191–202

Idrissou Y, Assani AS, Toukourou Y, Worogo HSS, Assogba BGC, Azalou M, Adjassin JS, Alabi CDA, Yabi JA, Alkoiret IT (2019) Systèmes d'élevage pastoraux et changement climatique en Afrique de l'Ouest: Etat des lieux et perspectives. Livest Res Rural Dev 31. http://www.lrrd.org/lrrd31/8/yadris31118.html

Idrissou Y, Assani Seidou A, Tossou MF, Baco MN, Sanni Worogo HS, Adjassin JS, Assogba BGC, Alkoiret Traoré I (2020) Perception du changement climatique par les éleveurs de bovins des zones tropicales sèche et subhumide du Bénin: Comparaison avec les données météorologiques. Cah Agric 29:9. https://doi.org/10.1051/cagri/2019032

IUCN (2010) Building climate change resilience for African livestock in sub-Saharan Africa – World Initiative for Sustainable Pastoralism, (WISP): a program of IUCN – The International Union for Conservation of Nature, Eastern and Southern Africa Regional Office, Nairobi, Mar 2010

Kaboré PN, Barbier B, Ouoba P, Kiema A, Some L, Ouedraogo A (2019) Perceptions du changement climatique, impacts environnementaux et stratégies endogènes d'adaptation par les producteurs du Centre-nord du Burkina Faso. VertigO- Rev Électronique En Sci Environ 19. https://doi.org/10.4000/vertigo.24637

Kiema A, Some L, Nacro BH, Compaore H, Kagone H, Kpoda CY, Bambara GT (2013) Stratégies d'adaptation des éleveurs de la zone Est du Burkina Faso aux effets des changements climatiques. Agron Afr 6:67–79

Kima SA, Okhimamhe AA, Kiema A, Zampaligre N, Sule I (2015) Adapting to the impacts of climate change in the sub-humid zone of Burkina Faso, West Africa: perceptions of agro-pastoralists. Pastoralism 5:16

Kosmowski F, Leblois A, Sultan B (2016) Perceptions of recent rainfall changes in Niger: a comparison between climate-sensitive and non-climate sensitive households. Clim Chang 135:227–241

Limantol AM, Keith BE, Azabre BA, Lennartz B (2016) Farmers' perception and adaptation practice to climate variability and change: a case study of the Vea catchment in Ghana. Springerplus 5:830

Mabe FN, Sienso G, Donkoh S (2014) Determinants of choice of climate change adaptation strategies in northern Ghana. Res Appl Econ 6:75–94

Mamba SF (2016) Factors influencing perception of climate variability and change among smallholder farmers in Swaziland. Indian J Nutr 3:138

McDonald JH (2009) Handbook of biological statistics. Sparky House Publishing, Baltimore

MEHU (2011) Deuxième Communication Nationale de la République du Bénin sur les Changements Climatiques. Cotonou Ministère L'Environnement L'Habitat L'Urbanisme

Nardone A, Ronchi B, Lacetera N, Ranieri MS, Bernabucci U (2010) Effects of climate changes on animal production and sustainability of livestock systems. Livest Sci 130:57–69. https://doi.org/10.1016/j.livsci.2010.02.011

Opiyo F, Wasonga OV, Nyangito MM, Mureithi SM, Obando J, Munang R (2016) Determinants of perceptions of climate change and adaptation among Turkana pastoralists in northwestern Kenya. Clim Dev 8:179–189

Oyekale AS (2014) Impacts of climate change on livestock husbandry and adaptation options in the arid Sahel belt of West Africa: evidence from a baseline survey. Asian J Anim Vet Adv 9:13–26

Polley HW, Briske DD, Morgan JA, Wolter K, Bailey DW, Brown JR (2013) Climate change and North American rangelands: trends, projections, and implications. Rangel Ecol Manag 66:493–511

Salack S, Muller B, Gaye AT, Hourdin F, Cisse N (2012) Multi-scale analyses of dry spells across Niger and Senegal. Sécheresse 23:3–13

Sanfo A, Sawadogo I, Kulo EA, Zampaligre N (2015) Perceptions and adaptation measures of crop farmers and agropastoralists in the eastern and plateau central regions of Burkina Faso, West Africa. FIRE J Sci Technol 3:286–298

Sanogo K, Binam J, Bayala J, Villamor GB, Kalinganire A, Dodiomon S (2017) Farmers' perceptions of climate change impacts on ecosystem services delivery of parklands in southern Mali. Agrofor Syst 91:345–361

Sanou CL, Tsado DN, Kiema A, Eichie JO, Okhimamhe AA (2018) Climate variability adaptation strategies: challenges to livestock mobility in south-eastern Burkina Faso. Open Access Libr J 5:1–17

Sautier M (2013) Outiller l'adaptation des élevages herbagers au changement climatique: de l'analyse de la vulnérabilité à la conception participative de systèmes d'élevage. Thèse de doctorat, Université de Toulouse

Uddin MN, Bokelmann W, Dunn ES (2017) Determinants of farmers' perception of climate change: a case study from the coastal region of Bangladesh. Am J Clim Chang 6:151

Zakari S, Tente BAH, Yabi I, Imorou IT, Tabou T, Afouda F, N'Bessa B (2015) Vulnérabilité des troupeaux transhumants aux mutations climatiques: analyse des perceptions et adaptations locales dans le bassin de la Sota à Malanville. Afr Sci Rev Int Sci Technol 11:211–228

Zampaligré N, Dossa LH, Schlecht E (2014) Climate change and variability: perception and adaptation strategies of pastoralists and agro-pastoralists across different zones of Burkina Faso. Reg Environ Chang 14:769–783

Zhang X, Yang F (2004) RClimDex 10 user manual climate research branch Environment Canada

8

Corona Virus, Climate Change and Food Security

Nkiru Theresa Meludu and Toyin Abolade

Contents

N. T. Meludu (✉) · T. Abolade
Department of Agricultural Economics and Extension Faculty of Agriculture, Nnamdi Azikiwe University, Awka, Nigeria
e-mail: nt.meludu@unizik.edu.ng

Abstract

Coronavirus disease of 2019 (COVID-19) is a current pandemic causing lockdown of cities and countries. The nature of this disease and the global cases are still considered as deadly all over the world. Analogous was drawn between the current COVID-19 pandemic and some of the other contemporary crises of the world as regards to climate change in addition to food shortage. Also, Survey Monkey instrument was used to generate emperical evidences from 514 respondents on covid-19 awareness and the effect on food security. Effects of diseases on climate change, such as the increasing frequency and strength of extreme weather events or the expanding range and spread of diseases was considered. Then, the relationship between the COVID-19 pandemic and climate change was investigated. Prior to the pandemic, climate change already had adverse effects on agriculture and vice versa, which led to food insecurity. The need for fruits as well as leafy and root vegetables in peri-urban and urban areas is increasing, as well as the food shortage. A drop in agricultural production will be expected in the future if the pandemic continues for a few more months. The perception and adherence to the preventive measures for this pandemic were determined to reduce its spread and lessen its effect on agricultural production as well as to improve food security.

Keywords

Covid-19 · Climate outcomes · Consequences · Agriculture · Food insecurity

Introduction

A total of 17 sustainable development goals (SDGs) have been set to transform our world, namely: no poverty, zero hunger, good health and well-being, quality education, gender equality, clean water and sanitation, affordable and clean energy, decent work and economic growth, industry, innovation and infrastucture, reduced inequality, sustainable cities and communities, responsible consumption and production, climate action, life below water, life on land, peace and justice strong institutions, partinership to achieve the goals (United Nations Department of Economic and Social Affairs (UNDESA) 2015).

SDGs are the organized principles for addressing environmental improvement objectives and sustaining the capacity of the usual structure to offer natural resources, flora and fauna services upon which the economy and humanity depends is crucial.

This chapter identified some of the issues influencing climate change and the effect of climate change on food security. It also discussed coronavirus disease 2019 (COVID-19), which is the current pandemic causing lockdown of the entire economy due to its global occurrence (World Health Organisation (WHO) 2020). Some emperical evidences were determined with the use of "Suervey Monkey" Instrument that generated responses from respondents from different parts different of Nigeria at the peak of the pandemic. This chapter highlighted the pandemic nature of COVID-19 and other hidden evidences in litreature. Essentially, some of the health constraints on climate change include the increasing occurrence of extreme weather events and the growing rate and spread of illnesses, such as malaria.

> **COVID-19,** a persistent respiratory disease caused by corona virus is transmitted chiefly by contact with human and infected surfaces and the individual shows symptom of fever, cough, and shortness of breath and may advance to pneumonia, respiratory failure and death.

Furthermore, the relationship between pandemic and climate change was investigated, although climate change already had adverse effects on agricultural production, which led to food insecurity, especially in under developed countries. It is expected that agricultural production will drop in the subsequent years if COVID-19 pandemic persists for more months. The vulnerable groups include small-scale farmers, migrants, labourers/workers, nomads, and fisher folks who have been prevented from working on their farms and from taking care of their animals/fisheries will fell the effect. These groups also face difficulty in obtaining access to proper marketplaces to retail their produces and processed goods as a result of increased food prices and inadequate purchasing resources. Informal laborers are also affected leading to hike in labor cost and income losses because of delayed harvesting of the produce.

Conceptualizing the Pandemic Nature of Coronavirus

COVID-19 could be described as a respiration set of severe acute respiratory syndrome coronavirus 2 (SARS-CoV-2) (World Health Organization (2020). This virus spreads primarily between people during close contact, often through small drops of discharge from the nostrils once an infested individual sneezes or coughs into the air. The droplets usually drop to the ground/surfaces without being suspended in the air moving 2-m distances (Nigerian Centre for Disease Control (NCDC) 2020). People may contract the virus by touching polluted surfaces and then touching their faces (mouth, nose, and eyes). The first case of COVID-19 occurred in December 2019 in Wuhan, China's Hubei Province. Then, it subsequently spread globally, which resulted in the ongoing 2019–2020 coronavirus pandemic (Hui et al. 2020; World Health Organization (WHO) 2020). COVID-19 has an incubation period of usually 2–14 days for people who develop symptoms; however, some people who are infected do not develop any form of symptoms and are thus called asymptomatic. While it is easy to isolate and treat infected people that develop symptoms, the asymptomatic patients cannot be easily identified, isolated, and treated, thus posing a great risk of spreading of the virus. This brings fear to the people and also has crippled many economies worldwide, affecting agricultural production, thus causing food scarcity, hunger, and food insecurity. Interestingly, researchers worldwide are working tirelessly to develop vaccines/drugs and treatment regimes of the disease.

Symptoms Associated with COVID-19 (NCDC and Hopkins University 2020; WHO 2020)

1. Cough – Coughing is a common impulse action that clears the throat of mucus or foreign irritants. While an individual coughs to free the throat at intervals, some amount of situations can lead more frequent coughing, which could be as a result of Covid-19.
2. Loss of appetite – This is referred to as anorexia and could be caused by a diversity of conditions and diseases. Some of these conditions could be temporal and rescindable, such as defeat of appetite or a side effect of medicine. Majority of the situations could be very severe, for instance, the consequences of disease causation (could be covid-19). Continuous absence of hunger should be evaluated by a healthcare professional.
3. Difficulty in breathing – This could be described as uneasiness and could be related to numerous diverse conditions. It is important to note that frequent incidents of shortness of breath could be a symptom of a severe illness (such as Corona virus) that needs attention.
4. High temperature/fever – The common body temperature is different for many people and changes during the day. If the body temperature is 38 °C, it is considered as a high fever. Several conditions could lead to fever, but it is usually caused by the body fighting an infection (may be covid-19).
5. Sneezing – This is a sudden reflex discharge of air from the nose and mouth due to irritation of the nostrils, which is caused by exasperation to the mucous membranes of the nose or throat. It may be very inconvenient and could be symptom of Corona virus.
6. Headache – Pain which occurs along the nerves, blood vessels, and muscles that is felt in the head. However, if it persists for 3 days, the doctor should be contacted because it could be an indication of an underlying disease.
7. Fatigue – This could be bodily, mental, or both, an indication which could be difficult to describe; however, words like tired, exhausted, and worn-out may be used.
8. Joint pains – This refers to the uneasiness and pain felt in some joints of the body, such as those of the shoulders, hips, elbows, and knees. Occasionally, joint pain is a result of an illness or injury.
9. Runny nose – This is an indication of several health situations resulting in the flow of secretion from the nostrils. This secretion protects material generated from the slimy tissue that lines the nasal cavity, which mixes with the saliva and drips down the back of the throat.

10. Sore throat – This is caused by a virus-related contamination, such as cold or influenza. It is also very important that symptoms are appropriately observed.

Preventive Measures of COVID-19

There are preventive measures recommended to prevent COVID-19 infection, such as self-isolation when feeling sick; avoiding busy spaces; practicing good hygiene; proper handwashing; using alcohol-based sanitizers; maintaining social and physical distancing from others, especially those with symptoms; covering mouth while coughing and sneezing with tissues, which should be thrown properly in a closed dustbin; and preventing from touching the face with unwashed hands. Furthermore, mask should be used outdoor, where additional non-pharmaceutical techniques cannot be adhered, to reduce the rate of transmission (NCDC 2020; WHO 2020).

Strategies Implemented by the Government Globally to Control and Prevent the Spread of COVID-19

1. Total and partial lockdown for weeks or months of states/regions affected by COVID-19
2. Enforcement of curfew between certain hours of the day to prevent people from going out and contracting the disease
3. Closure of the country/state borders (land, air, and waterways) to prevent further spread of the disease
4. Banning of the interstate movement
5. Fumigation of public places with disinfectants (marketplaces, parking areas, hospitals, congested areas, roads, etc.)
6. Enforcement of the use of surgical gloves, medical masks for healthcare givers, and fabric masks for non-medical personnel
7. Restriction of people's movements to only essential services (healthcare workers, security personnel, and journalists)
8. Opening of markets for some days of the week
9. Distribution of palliatives to the poorest of the poor in the country to enable citizens to fully comply with the total lockdown

However, some of the measures implemented to prevent the spread of the virus, such as fumigation of public spaces with chemicals and the use of personal protective equipment (PPE), which are not properly disposed, can lead to land, water, and air pollution, which will later negatively affect the environment, possibly resulting in environmental challenges and food insecurity and these could be worst in less-developed countries.

Diagnosis of COVID-19

The seriousness of the pandemic differs from person to person. COVID-19 infection could show minor sypmtoms, such as ordinary flue or common cold, or no symptoms at all. As the infection progresses, the symptoms change based on gender, generational, and the type of underlying illness. People with mild symptoms of COVID-19 infection usually get better in 14 days or less, although it takes 21 to 30 days to achieve full recovery (NCDC 2020; WHO 2020). Older people or those with underlying diseases, such as hypertension, diabetes, cardiovascular diseases, chronic respiratory diseases, and cancers of any type, are susceptible to this infection (NCDC 2020), while children experience less severe symptoms. The mortality rate of adults less than 50 years who were infected with COVID-19 is less than 0.5%. Conversely, those older than 70 years had the highest mortality rate from the disease. The major problems related with COVID-19 include sepsis, excessive clotting, and damage to the vital organs in the body. Up to 30% of the individuals infected with COVID-19 exhibited elevated liver enzymes, known as *transaminases* (Sanders et al. 2020), resulting in organ failiure and eventually death.

Transformation of Situations Causing Typical Climate Change

Climate change is described as variation on usual climate conditions of an area across periods due to natural circumstances or artificial activities causing degradation of the environment (Nwosu 2012). Climate change is considered as a long-term shift in the average weather situation of a region, such as its typical temperature, rainfall, and wind patterns. This variation in weather indicates that the consequences estimated in a lot of countries will vary over the coming decades.

> Weather variations are global phenomenon of climate transformation characterized through fluctuations observed on weather of the planet regarding temperature and precipitation, which are especially instigated through human activities (Food and Agriculture Organisation (FAO) 2002). The balancing of the weather condition to sustain ecosystems is under threat, as well as the future of humanity and the steadiness of the world economy.

In other words, climate change refers to the deviation in the geometric distribution of the average weather circumstances over a prolonged period of time. Climate change induces an increase in the occurrence of extreme weather events. The characteristics of weather variation poses risks to agricultural sustainability and results in food shortage, high level of poverty, and low level of human and physical development. Uninterestingly, weather variations negatively affect human health, water resources, land use, coastal infrastructure, and the environment (International Fund for Agricultural Development (IFAD) 2010), particularly in developing countries, such as Nigeria, where irrigation is not often practiced, as well as in regions that experience heavy rainfall, thus causing extreme flooding and leaching.

Thus, the apprehension of weather variation cannot be overemphasized which is based on the biggest problem facing the world, affecting everything and everyone, due to the fact that it plays an important environmental influence on ecosystem. Climate change affects the ecosystems in numerous ways, and intermingles with poor growth. The cumulative impacts could lead to intense environmental variations (United States Global Change Research Program (USGCRP) 2009). The severity and frequency of diseases are often related with unfavorable weather conditions and unpredicted health concerns spreading in regions where they have not occurred before.

> The consequences of weather variation on the availability, accessibility and stability of food supplies can cause major alterations in people's diet choices, which in turn have consequences on the health of the households and low recovery rate from the pandemic.

Nelson et al. (2018) has reported that climate change is expected to alter global nutrient availability, thereby exerting aftermath effects on malnutrition. The increasing CO_2 levels are also expected to reduce the levels of proteins, iron, zinc, and vitamins B1, B2, B5, and B9 in staple foods and docosahexaenoic acid content of fish, which will cause malnutrition to millions of people worldwide after the COVID-19 pandemic (Colombo et al. 2019). COVID-19 pandemic and devastating consequences, and how to stop its spread have become the focus of the world. The greenhouse gases (GHGs) that significantly contribute to climate change, include carbon dioxide, methane, nitrous oxide, ammonia, and other trace gases (Intergovernmental Panel on Climate Change (IPCC) 2007). The other factors causing weather variations are natural causes and human activities. Human activities, such as fumigation and disinfection of public places and improper disposal of used PPEs during the COVID-19 pandemic, also immensely contribute to climate change. The scorching of fossil fuels and the conversion of land for forestry to other uses, including agriculture, are also examples of human activities. Conversely, the natural causes of climate change include variations in ocean currents and atmospheric circulation.

Effects of Weather Variation on the Environment and Agriculture in Nigeria

The effects of weather variation on food production differ from one country to another which is dependent on the current climate condition affecting agricultural productivity differently. Nigeria, like all the other countries in the sub-Saharan Africa, is highly vulnerable to the impact of climate change. Climate change variations from 1960 and 1999 in parts of Nigeria presented visible incidences of deficiency for a cumulative period of approximately 8–18 years in some parts of Nigeria (Obioha 2009; Muhammed et al. 2011). Uncertainties about the farming season due to the changes in rainfall pattern resulted in crop failure, thus leading to food shortages because of poor harvests. An increase in rainfall is also favorable for the outbreak of pests and crop diseases, which is detrimental to crop production. Specifically, pests and crop diseases are visible in response to weather variation, which hypothetically threaten livestock in the drier northern states of Nigeria (Food and Agriculture Organisation (FAO) 2002). Small-scale farmers face the problem of flooding, which destroys the infrastructure used to store or transport food from production areas to markets, thereby discouraging the farmers to produce more food (Aluko et al. 2008). Global warming also negatively affects the seafood and aquatic lives that are important to humans. In addition, climate change affects not only food security but also the livelihood of the people, thus obstructing the advancement of many developing and underdeveloped countries. It is quite obvious that the most serious challenge to agricultural production and mankind worldwide, especially in Africa, is climate change.

Effects of Climate Change on Food Security

African countries that rely solely on natural resources and rainfed agriculture are more vulnerable to the risks of climate change (IFAD 2010). The consequences of climate change effects are irreversible, potentially disastrous to the entire nation, most especially the developing and underdeveloped nations. For example, it was estimated that by 2020, agricultural production would decline by 50% in some countries with rainfed agriculture (Nanki et al. 2010), and this will be worsened by the COVID-19 pandemic. The Intergovernmental Panel on Climate Change (IPCC) in 2008 analyzed that a temperature increase of 2.5 °C would result in a loss of 1.0–1.5% of the gross domestic product (GDP) in developed countries and 2–9% in developing countries. Generally, climate change causes floods, droughts, or tropical storms, which can take a major toll on a country's economy if a significant part of economic activity is sensitive to the weather and climate. Therefore, if the current trend of GHG emissions continues, the per capita GDP loss by 2022 will be as much as 3–35% (Tamirisa 2008; Harris et al. 2017). Africa is a good example of the influence of climate variability on a developing country's economy (USAID 2010; Zenebe et al. 2011). Boko et al. (2007) and IFAD (2010) reported that the rainfed agriculture and pastoral communities in Africa are dependent on annual rain and were identified to be more susceptible to the effects of climate change. Climate change have negative impacts on food production, especially livestock, in terms of CO_2 and methane emissions. Livestock uses about 32% of arable land, and cattle rearing causes up to 70% of land degradation and pollution. Therefore, if forests are lost, it will be difficult to fight climate change (Meludu 2019). The sheer quantity of animals being raised for human consumption is a threat to climate sustainability. Another scenario is that 1 ha of land for beef production will feed one person, the same hectare for rice production will feed 19 persons, and the same hectare for potato production will feed 22 persons. This competition is too high to achieve food security and eradicate poverty due to degradation and pollution.

Concepts of Food Security

Food security exists at all periods with individuals having financial access to necessary, harmless, and nutritious food that provides adequate requirement for sustainable life expectancy (FAO 2002; Meludu 2009; Meludu and Ewebiyi 2010). Food security and food insecurity are considered to be dynamics that depend solely on causes of not having enough food and the management. However, the partial and total lockdown implemented worldwide due to the COVID-19 pandemic has affected numerous economic activities, such as food distribution, restaurants, hospitality industries, and transportation activities. During the COVID-19 pandemic, many countries closed their borders, preventing the free flow of goods in to different countries and even within the cities. This restriction caused constraints in food distribution and food hoarding, which posed great challenges to food security. This is especially to the vulnerable population including; aged people (senior citizens), children, pregnant women and those with immune-compromised status. Due to the pandemic, food prices have increased worldwide; staple foods such as wheat, *garri*, rice and beans, tubers, meats, fruits, and vegetables became expensive. Adequate and effective support to encourage unhindered transportation of food items from source locations to various destinations, home deliveries became necessary to reduce human interactions and prevent the possibility of getting infected. When food security is compromised, the consequence is food insecurity.

Food accessibility is affected by the production, distribution, and stability of food (Fig. 1). Food accessibility refers to the easy access to food that is essential to combat hunger and overcome food insecurity. An individual or household's access to food can be influenced by their income and health conditions. Consumers must be able to have the purchasing power in order to have access to food. Due to the COVID-19 pandemic, the rate of unemployment has increased, which resulted in food insecurity. Individuals need to eat sufficiently to develop strong immunity against the virus and to be physically healthy to be able to digest and utilize the nutrients in the food consumed. Food utilization can also be influenced by nutritional values, health status of an individual, safety of the food, and how it is prepared before consumption. Food will be well utilized when people strictly adhere to the important safety precautions against COVID-19. Food stability is guaranteed if the governments worldwide make adequate preparations to reduce the rate of

Fig. 1 Dimension of food security. (Source: Francesco et al. (2011))

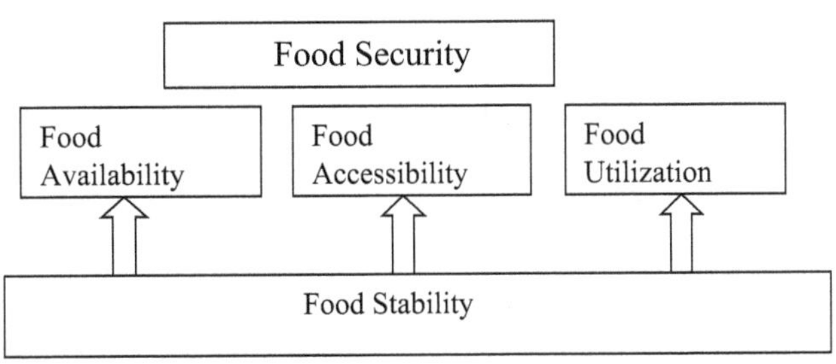

postharvest losses to make food available during this pandemic so that people will remain healthy and safe.

Therefore, to achieve all the highlighted dimensions of food insecurity, the pandemic should be dealt with accordingly to reduce its impacts on climate change so as to avoid food shortage. If adequately managed, the resultant effects will reduce the occurrence of food insecurity globally.

Food Insecurity

Food insecurity could be considered as deficiency in access to adequate food amount in terms of quality and quantity. Inadequate diet and reduced immune system amplify vulnerability to diseases (Meludu 2011). Low diet is correlated with the increasing development of non-communicable diseases (NCDs) (Velmurugan et al. 2017).

Climate change-related incidents recurring everyday requires immediate action (Solly 2019). Therefore, a scientific innovation strategy for mitigating the impacts of climate chnage on agricultural systems is important. Increased food traceability aimed at preventing food loss due to contamination issues, prediction of extreme events is very germain at this period of pendemic (Dhaliwal and Williams 2019; Fabregas et al. 2018; Nkurunziza et al. 2019).

Food contamination caused by climate change also threatens the achievement of the SDGs, namely SDG 2 (zero hunger), SDG 3 (good health and well-being), SDG 8 (decent work and economic growth), SDG 12 (responsible consumption and production), and SDG 13 (climate action) and indirect effect on the other SDGs. Labor scarcities impacted producers, processors, traders, and distribution companies in food distribution chains necessitate workforces to be in near vicinity. The United Nations World Food Programme (UNWFP) predicted that approximately 265 million people will experience increased food insecurity by the end of 2020.

> Food insecurity can be illustrated by durations such as; chronic food insecurity is a long-standing food scarcity. Temporary food insecurity is a momentary circumstance of food insecurity where a populace agonizes from food insecurity transition at the time there is an unexpected drop in the capability to produce food for adequate nutritional status and lastly, Seasonal food insecurity is a condition that could happen from time to time.

Food insecurity "hot spots" according to the UNWFP include the following:

- Substantial conflict-affected states, where food supplies are problematic due to the pandemic. The Fulani nomads are also moving due to climate change which is destroying crops and agricultural land. Uninterestingly, many farmers and their families are displaced from their communities and killed
- Countries experiencing numerous emergencies due to climate change consequences that affect food production worldwide, accompanied by the high incidence of flood preventing farmers from going to their farms to plant

- Vulnerable individuals constantly experiencing food insecurity even before the pandemic
- Countries with significant currency depreciation and other product price failure (dropping the volume to import food)

Effects of the Coronavirus 2019 (COVID-19) Pandemic and Climate Change on Food Security

The COVID-19 pandemic has negative effects on climate change which lead to food insecurity. The process of fumigating public places (markets, churches, parking areas, offices, and the entire environment) pollutes the air which eventually affects the weather condition (Fig. 2).

COVID-19 has caused a dramatic loss of human lives worldwide. It also presented an unprecedented challenge with deep social and economic consequences, including food insecurity and malnutrition. An impact on the crusade of labor force and the distribution of inputs has also caused unacceptable constraint to food production. However, agriculture and food-related logistic services was considered to be among the important concerns during the COVID-19 pandemic; thus, food production and distribution should not be compromised.

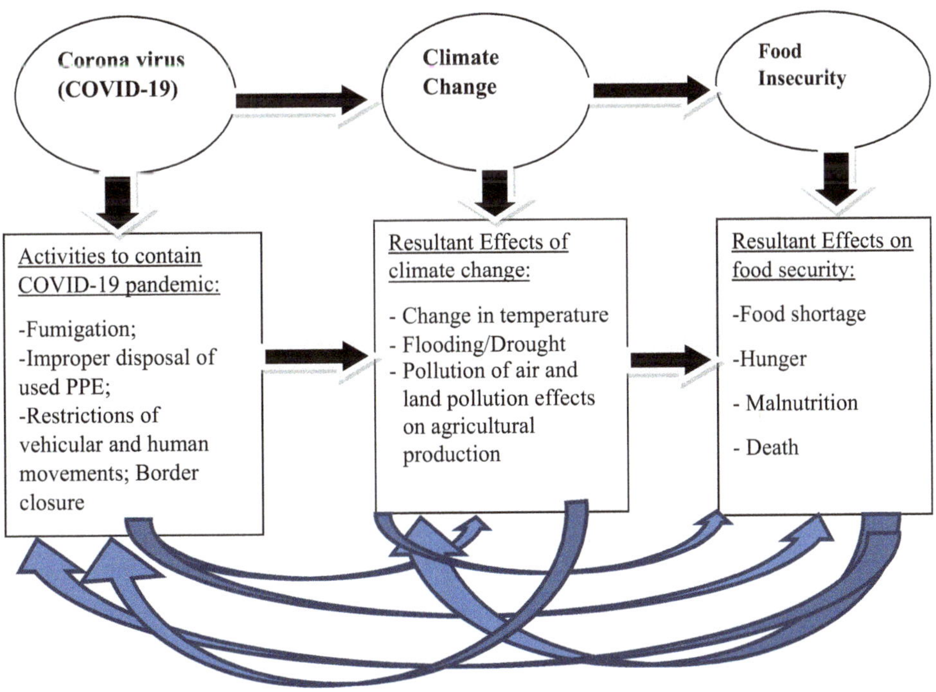

Fig. 2 Conceptual framework of the interface between COVID-19, climate change, and food security

The government in every country adopted the idea of lockdown as part of the precautionary measures to stop the spread of COVID-19 among its citizens. On the contrary, this lockdown resulted in food shortage in that farmers, marketers, processors, and other stakeholders in the value/supply chain were unable to carry out their enterprise characteristics and the global food system has been destroyed.

Empirical Evidences Obtained Through Online Survey on COVID-19 Awareness and the Effect on Food Security

This study used SurveyMonkey (Monkey Survey 2020) to obtain answers to selected questions that were raised with regard to the effects of the COVID-19 pandemic on farming and food security. This study included 514 agricultural stakeholders as participants, of which 55.6% are male and 44.4% are female. It is so disappointing to note that many of these participants (45%) were not taking COVID-19 very seriously, and some were even unsure whether this disease is real or not. They considered it as politicians or rich man disease. This resulted in non-adherence to the safety measures implemented by the government, thereby causing the rapid spread of the disease. About 60% of the respondents learned of this pandemic through radio, television, and Internet, 15% from healthcare workers, 15% from family members, 5% from market places, and 5% from workplaces. This implies that many farmers did not have access to information and communications technologies (Fig. 3) or regular power supply. Thus, only a few people who had sufficient information were able to take adequate measures. This will cause many people to contract the virus, thereby leading to death and affecting agricultural activities which causes food insecurity.

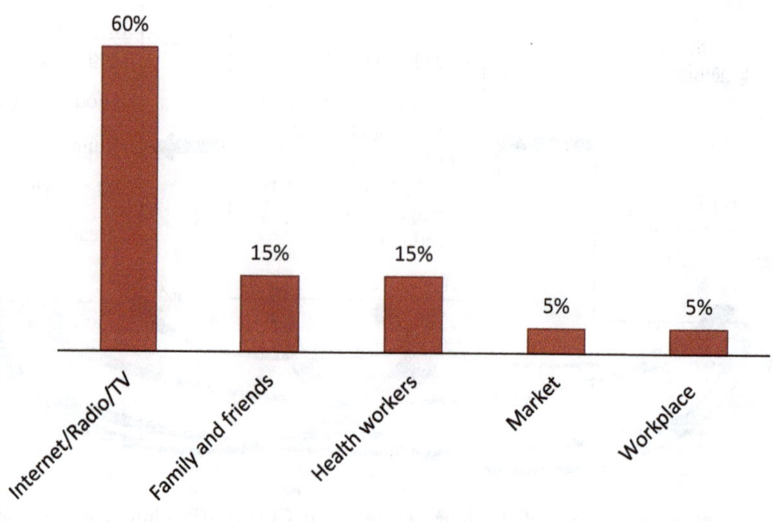

Fig. 3 Channel of information on the COVID-19 pandemic

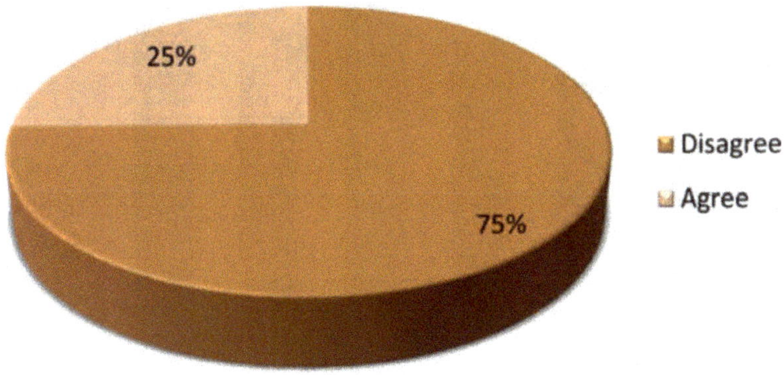

Fig. 4 Farmers' access to palliatives

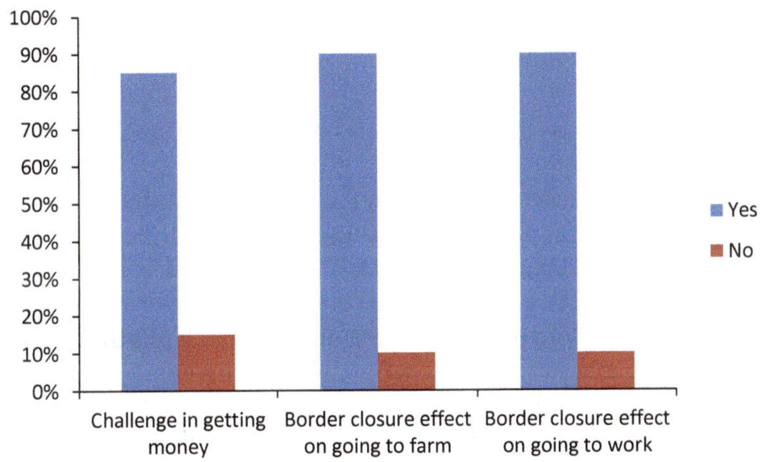

Fig. 5 Effect of COVID-19 on access to money, farm, and office work

During the period of the lockdown to contain COVID-19, when all economic activities were put on hold, many industries could not pay the salaries of their workers; thus, there was either a pay cut or staff layoff. Due to this, the purchasing power was affected and food accessibility became difficult to achieve. Majority (75%) of the respondents disagreed with the way the government handled the palliative distribution as not everyone was given, especially the farmers, whereas less than average (25%) agreed (Fig. 4).

Farmers that were supposed to be on the farm working were asked to stay home as one of the measures to prevent them from contracting the disease. Almost all (90%) the farmers adhered to this, and 85% did not have access to money (Fig. 5). Most crop and livestock farmers in the developing countries rely solely on rainfed agriculture, whereas only few depend on irrigation. Unfortunately, the peak of COVID-19 occurred at the beginning of the rainy season, which is in April. The farmers were unable to perform their seasonal farming activities. Eventually, this will result in food shortage and food insecurity. Transport route blockage hinders the

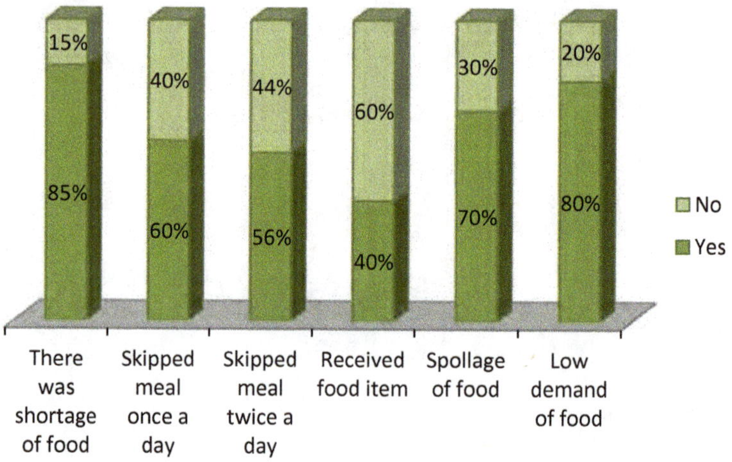

Fig. 6 Distribution of the shortage of food, skipping of meals, food gift, spoilage of agricultural produce, and low demand for food

distribution of fresh goods; thus, high levels of postharvest loss and waste may be experienced.

Marketers play significant roles in ensuring adequate distribution of food products. During the pandemic, marketers were also asked to stay home for a period of weeks or months, depending on the number of index cases recorded in the region. During this period, marketing activities were disrupted as marketers could not buy or sell food as expected. Moreover, there were no available buses/trucks to transport goods and services; interstate borders were also closed. During this, food accessibility was a major challenge for households, thus, eventually leading to food insecurity. Processors are another category of people involve in food distribution. Due to lockdown, the raw materials used for food processing were not available. Therefore, much of foods processing/income were not achieved. This invariably resulted in food shortage and food insecurity and resulted in many (85%) households not having access to food. A high percentage of households reported skipping food once (60%) or twice (56%) a day (Fig. 6). Some crops are supposed to be harvested before the rain in flood-prone areas. However, due to the pandemic, they were left on the ground to spoil (70%). These measures also resulted in labor shortage which disrupted the production and processing of food, most especially for labor-intensive crops. Migrant workers left to their states, and those who stayed were not allowed to come out, which led to high labor cost even after the relaxation of some measures.

Conclusion

Climate variation causes events that affect many economies and has been linked to poor food production even before the pandemic. Thus, more effects were felt as part of the pandemic consequences, which led to food shortages. Lockdown and closure of borders led to farm labor shortage and increase in food prices. In addition, the amount of food that would have been produced by farmers reduced. Crops harvested by the farmers and ready for sale got wasted on the farm due to the movement restrictions. There were also few trucks available for transporting produces and food products, which led to an increase in transportation costs. Many low-income households, especially in the rural areas, could not afford to buy enough food to sustain the family, thereby causing hunger and malnutrition and lowering their immunity, which made them susceptible to diseases. This pandemic has affected the sustainable development goals toward achieving food security, namely, goal 1: no poverty, goal 2: zero hunger, goal 3: good health and well-being, goal 4: quality education, goal 5: gender equality, goal 6: clean water and sanitation, and goal 7: achieving human development goals and sustaining the ability of natural system to provide natural resources and ecosystem services by 2030.

Recommendation

Timely and effective policies should be implemented to prevent more effects on climate variations and also to reduce the consequences of COVID-19 so that extreme poverty and hunger can be avoided. To overcome challenges on food price, an adequate price control mechanism and a price monitoring system for food produce must be in place to evaluate the actions. Quality supervision of food items by the appropriate organs of the government is very important in ensuring food safety and quality, especially during and after the pandemic. Countries are encouraged to produce their own agricultural products. People are now advised to engage in agriculture to reduce the effect of the pandemic on food security.

References

Aluko AP, Oyeleye B, Sulaiman ON, Ukpe IE (2008) Climate change: a threat to food security and environmental protection. In: Proceedings of the 32nd annual conference on forestry Association of Nigeria

Boko M, Niang I, Nyong A, Vogel C, Githeko A, Medany M, Osman-Elasha B, Tabo R, Yanda P (2007) Climate Change 2007: impacts, adaptation, and vulnerability. In: Parry ML, Canziani OF, Palutikof JP, van der Linden PJ, Hanson CE (eds) Contribution of working group II to the fourth assessment report of the Intergovernmental Panel on Climate Change. Cambridge University Press, Cambridge, UK, pp 433–467

Colombo SM, Rodgers TFM, Diamond ML, Bazinet RP, Arts MT (2019) Projected declines in global DHA availability for human consumption as a result of global warming. Ambio 37:1–16

Dhaliwal DS, Williams MM (2019) Optimum plant density for crowding stress tolerant processing sweet corn. PLoS One 14(9):e0223107. [online]. Accesses 25 May 2020

Fabregas N, Lozano-Elena F, Blasco-Escamez D, Tohge T, Martinez-Andujar C, Albacete A, Osorio S, Bustamante M, Riechmann JL, Nomura T, Yokota T, Conesa A, Alfocea FP, Fernie AR, Canno-Delgado AI (2018) Over expression of the vascular brassinosteroid receptor BRL3 confers drought resistance without penalizing plant growth. Nat Commun 9(1):4680

FAO (2002) Report of the world food summit five years after. FAO, Rome

Francesco B, Jessica F, Emile F (2011) The role of food and nutrition system approach in tackling hidden hunger. Int J Environ Res Public Health 8:358–373

Harris JM, Roach B, Anne-Marie C (2017) The economics of global climate change. A GDAE Teaching Module on Social and Environmental Issues in Economics. Global Development and Environment Institute, Tufts University Somerville

Hui DS, Azhar E, Madani TA, Ntoumi F, Kock R, Dar O (2020) The continuing 2019-nCoV epidemic threat of novel corona viruses to global health. Int J Infect Dis 91:264–266

IFAD (2010) Livestock and climate change Livestock. The International Fund for Agricultural Development (IFAD) Livestock Thematic Papers. http://www.ifad.org/lrkm/factsheet/cc.pdf. Accessed 9 Aug 2013

IPCC (2007) Summary for policymakers. In: Solomon S, Qin D, Manning M, Chen Z, Marquis M, Averyt KB, Tignor M, Miller HL (eds) Climate change 2007: The physical science basis. Contribution of working group I to the fourth assessment report of the Intergovernmental Panel on Climate Change. Cambridge University Press, Cambridge, UK/New York

Meludu NT (2009) Food insecurity and nutritional status of children in rural households (A case of Iddo LGA, Oyo State). Niger J Soc Work Educ 8:40–60

Meludu NT (2011) Farmers' awareness and utilization of disaster management strategies and training needs for sustainable food security and livelihoods in Nigeria. In: Brebbia CA, Kassab AJ, Divo EA (eds) Disaster management and human health risk II: reducing risk, improving outcomes. WIT Press, Southampton/Boston, pp 303–319

Meludu NT (2019) Communicating knowledge of climate change in achieving zero hunger – sustainable development goal (SDG). Paper presented at the one day media round table discussion to the University on Communicating Knowledge on Climate Change Information for Sustainable Decision Making in Nigeria. By Centre for Sustainable Development Nnamdi Azikiwe University, Awka May, pp 1–4

Meludu NT, Ewebiyi IO (2010) Policy development and opportunities in ICT for sustainable food and nutrition security in Nigeria. Afr J Psychol Stud Soc Issues 13(1):137–151

Monkey Survey (2020) Survey Monkey, San Mateo, pp 1–23. http://www.surveymonkey.com. Accessed 5 May 2020

Muhammed G, Jaliye MM, Usman S, Baniwa E (2011, March) Adaptive strategies to climate change variability and change in rural communities. In: Proceedings of the 16th annual national conference of agricultural extension of Nigeria (AESON)

Nanki K, Million G, Beneberu S, Zegeye T, Gebeyehu S, Endale A (2010) Adapting to climate change in the water sector: assessing the effectiveness of planned adaptation interventions in reducing local level vulnerability. RiPPLE. Water-Aid-Ethiopia, Addis Ababa

NCDC (2020) Covid-19 Nigeria. Available at ncdc.gov.ng. Accessed 7 June 2020

Nelson G, Bogard J, Lividini K, Arsenault J, Riley M, Sulser TB, Mason-D'Croz D, Power B, Gustafson D, Herrero M, Wiebe K, Cooper K, Remans R, Rosegrant M (2018) Income growth and climate change effects on global nutrition security to mid-century. Nat Sustain 1(12):773–778

Nkurunziza L, Chirinda N, Lana M, Sommer R, Karanja S, Rao I, Romero Sanchez MA, Quintero M, Kuyah S, Lewu F, Joel A, Nyamadzawo G, Smucker A (2019) The potential benefits and trade-offs of using sub-surface water retention technology on coarse-textured soils: impacts of water and nutrient saving on maize production and soil carbon sequestration. Front Sustain Food Syst 3. https://doi.org/10.3389/fsufs.2019.00071

Nwosu LE (2012, March) Climate change and food security: implication for effective agricultural extension services in Nigeria. In: Proceedings of the 17th annual national conference of agricultural extension of Nigeria (AESON)

Obioha EE (2009) Climate variability, environmental change and food security nexus in Nigeria. J Hum Ecol 26(2):107–121

Oliva R, Ji C, Atienza-Grande G, Huguet-Tapia JC, Perez-Quintero A, Li T, Eom J-S, Li C, Nguyen H, Liu B, Auguy F, Sciallano C, Luu VT, Dossa GS, Cunnac S, Schmidt SM, Slamet-Loedin IH, Vera Cruz C, Szurek B, Frommer WB, White FF, Yang B (2019) Broad-spectrum resistance to bacterial blight in rice using genome editing. Nat Biotechnol 37(11):1344–1350

Report of WHO-China Joint Mission on Corona Virus Disease 2019 (COVID-19). 16–24 February 2020. Accessed 3 May 2020

Sanders JM, Monogue ML, Jodlowski TZ, Cutrell JB (2020) Pharmacologic treatment for Corona virus diseases 2019. A Review. PMID: 32282022. https://doi.org/10.1001/jama.2020.6019

Solly M (2019) One climate crisis disaster occurs every week, UN official warns. Smithsonian Magazine, 8 July 2019. https://www.smithsonianmag.com/smart-news/un-expert-warnsone-climate-crisis-disaster-occurs-every-week-180972573

Tamirisa N (2008, March) Climate change and the economy. Finance & Development. http://www.imf.org/external/pubs/ft/fandd/2008/03/pdf/tamirisa.pdf. Accessed 23 Apr 2020

United Nations Department of Economic and Social Affairs (2015) Sustainable development goals. Available at http://www.un.org/sustainabledevelopmentmaterials. Accessed 21 May 2020

United States Global Change Research Program (USGCRP) (2009) Global climate change impacts in the United States. In: Karl TR, Melillo JM, Peterson TC (eds) United States global change research program. Cambridge University Press, New York

USAID (2010) Adapting to climate variability and change a guidance manual for development planning. http://pdf.usaid.gov/pdf_docs/PNADJ990.pdf. Accessed 9 Apr 2020

Velmurugan, G., Ramprasath, T., Gilles, M., Swaminathan, K. And Ramasamy, S. (2017). Gut microbiota, endocrine-disrupting chemicals, and the diabetes epidemic. Trends Endocrinol Metabol, 28(8): 612–625. https://doi.org/10.1016/j.tem.2017.05.001

World Health Organization (2020) WHO Director-General's opening remarks at the media Briefing on COVID-19 (Press release) 11 March 2020. "Advice for Public"

Zenebe G, Stage J, Alemu M, Atlaw (2011) Climate change and the Ethiopian economy. A computable general equilibrium analysis: environment for development discussion paper series October 2011, EfD DP 11-09. http://www.rff.org/RFF/Documents/EfD-DP-11-09.pdf http://www.vanguard.com/2020/04/covid-19-maintaining-food-security-in-crisis-situations. Accessed 9 Aug 2019

9

Impacts of Environmental Change on Fish Production in Egypt and Nigeria: Technical Characteristics and Practice

M. L. Adeleke, D. Al-Kenawy, A. M. Nasr-Allah, M. Dickson and Desalegn Ayal

Contents

M. L. Adeleke (✉)
Department of Fisheries and Aquaculture Technology, The Federal University of Technology, Akure, Akure, Nigeria
e-mail: mladeleke@futa.edu.ng; mosunmolalydia@yahoo.com

D. Al-Kenawy · A. M. Nasr-Allah · M. Dickson
WorldFish, Abbassa, Abou-Hammad, Sharkia, Egypt

D. Ayal
Center for Food Security Studies, College of Development Studies, Addis Ababa University, Addis Ababa, Ethiopia

Abstract

A survey approach was applied to examine the technical characteristics of fish farming practice in Egypt and Nigeria. Critical issues such as floods and other vices were considered in bringing out the inference and level of aquaculture in both countries. Multistage sampling technique was used to select the study area and the number of respondents. Both primary and secondary data were used in the analysis. Eighty fish farmers were randomly selected from each of the country, making a total of 160 respondents. The socioeconomic characteristics revealed that aquaculture is an antique venture in Africa with Egypt taking the lead, i.e., 99% of the respondents practice in large-scales production of more than 11 Fadden/acre per, and 100% males depend mainly on agricultural drainage water for their earthen ponds. Ninety-eight percent cultured tilapia (*Oreochromis niloticus*) which was believed to have originated from the Nile River. In terms of financial performance and partial economic analysis, tilapia production commands more sales in Egypt, while catfish (*Clarias*) production is seen as a promising venture in Nigeria. The fish farmers have various perceptions and reactions toward environmental changes factors such as cost of labor/manpower and inputs, poor water quality, and fish extension services, and climate change impedes aquaculture development in the countries. To achieve the scale of aquaculture expansion as observed in Asian and other developed part of the world, efforts should be geared toward continental and regional integration in order to encourage aquaculture practices in Nigeria and other parts of Africa. More so, government intervention and incentives should be paramount in Egypt to reduce the excessive exploitation of the private input suppliers.

Keywords

Technical · Environmental · Fadden · Tilapia and catfish

Introduction

Aquaculture is an organized process in which the cultivation of aquatic animals and plants in fresh, brackish, and marine water is made, to increase fish production above the level that would be produced naturally (FAO 2016). Fish is the most-traded food commodities worldwide; more than 3.1 billion people depend on it for their average per capita intake of animal protein which is about 20%. Aquaculture has increased the world per capita fish supply to 20 kg in 2014 (FAO 2016). Global fish forecast model predicts that the share percentage of capture fisheries is expected to fall by half in 2030 compared to the 60% global production in 2011 after growing only by 2.8 million. Aquaculture is expected to grow by 30 million tons over this same period. In terms of food fish production, the model predicts that aquaculture will contribute 62% of the global supply by 2030 (FAO 2016). The problem of feeding the more than 9 billion people in the world by 2050 in the context of climate change and uncertainty (World Bank 2013) will be rightly addressed if improved aquaculture management is adhered.

Aquaculture seems to be multilocational and isolated. The Egyptians were probably the first in the world to culture fish as far back as 2500 B.C. from Japanese Resource Council, Science and Technology Agency as cited in Jhingran (1987). Aquaculture production in Egypt far exceeds that of the rest of the other Africa countries. It is a 2.2 billion USD in 2015. Egyptian aquaculture currently provides almost 79% of the country's fish needs, with almost all the output coming from small- and medium-sized privately owned farms (GAINS 2015).

Aquaculture was introduced into Nigeria in the 1950s. Recently, it is one of the fastest growing agricultural enterprises in the country. It impacted the nation's economy in terms of contributing over 1 million metric tons to Gross Domestic Product (GDP). It is best alternatives to meet protein demand; it is a business with brighter opportunities. Despite the fact that aquaculture is an age-old practice in some regions of the world, it is relatively new in most African countries and is significant to the economic, livelihood, and nutritional welfare of the populace. From all indications, Egypt has the largest aquaculture industry in Africa (Sumaila et al. 2014).

Therefore, the research questions, what are the technical factors to be considered in order to shrink the disparities? And environmental changes impact on fish production in Africa. This chapter classically, therefore, examines the technical characteristics of fish farming (Aquaculture) practice and impacts of environmental change on fish production in Egypt and Nigeria. Specifically, the socioeconomics characteristics of the respondents were determined using Kobocollect, fish farming system and operation were examined, and critical issues and institutional framework were considered bearing in mind the fish farmers' awareness on environmental changes.

Methodology

The Study Area

Egypt (see Fig. 1) is located between 30° 06′ N and 31° 25′ E. Based on the US estimates, the current population of Egypt is 94,660,721 which is equivalent to 1.27% of the total world population (Worldometers 2017), and the total land area is 995,560 km^2 (384,388 sq. mi). The climate is arid where most of the rains fall in the winter months with an average of 2–5 mm (0.1–0.2 in)/year with the exception of the Northern coast which sometimes can be as high as 410 mm (16.1 in) between the period of October and March (Worldometers 2017). The temperature range from 49.1 °F to 73.4 °F in low wind and sometimes average high temperatures vary from 62.6 °F in wintertime to 89.6 °F in summertime were recorded.

The Nile is the primary water source of Egypt and provides more than 95% of all water available to the country. Over 60 million people are connected with the agricultural sector which constitutes 20% of GDP and consumes about 80% of the

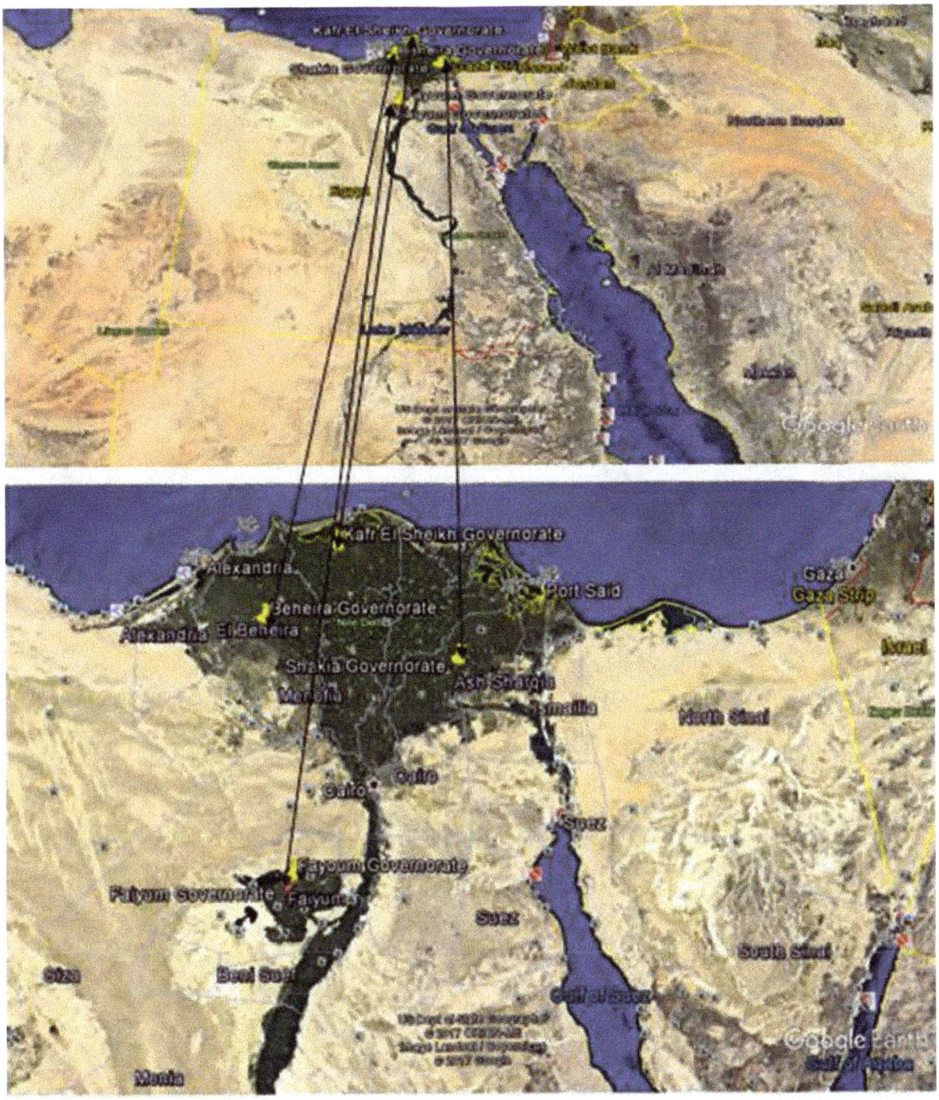

Fig. 1 Google Earth Map showing the study area in Egypt. (Source: Authors creation, 2017)

water budget. Aquaculture in Egypt is supported by water from irrigated agriculture and lakes such as lakes Burullus, Manzala, and Edku all located in Egypt.

Nigeria (see Fig. 2) lies exclusively within the tropical zone in West Africa. It borders with the Republic of Benin in the west, Chad and Cameroon in the East, and Niger in the North. The current population is 190,423,532 which is equivalent to 2.55% of the total world population and ranked 7th in the list (Worldometers 2017). The total land area is 910,802 km^2 (351,662 sq. mi) with 9° 82′ 00″ N; 8° 67′ 530″ E coordinates (see Fig. 2). The northeast and southwest winds are the two principal wind currents that make Nigeria to be distinguished with two distinct seasons: a wet season from April to October, with generally lower

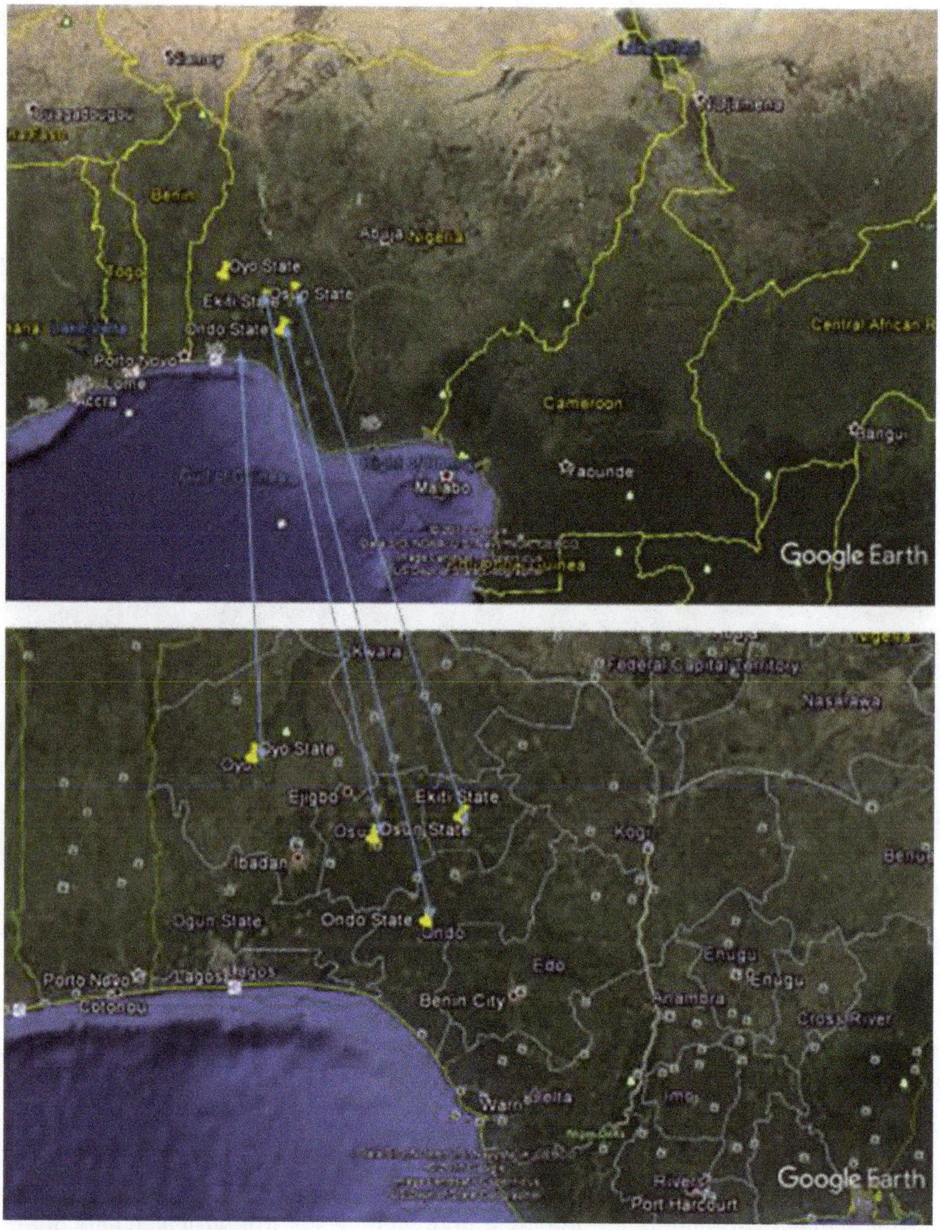

Fig. 2 Google Earth Map showing the study area in Nigeria. (Source: Authors creation, 2017)

temperatures, cloudy and rainy weather upshot of the southwest wind current. There are wide climatic variations in the country depending on the season and region. Nigeria got her name from the river, Niger River. Nigeria is indeed blessed with marine, brackish, and freshwater in addition to lakes, rivers, deltas, reservoirs, dams, rain, floodplains, inland, and coastal waters to support fisheries and aquaculture in the country.

The Design

In this research, explanatory mixed research design was applied to address specific objectives of the research work. The selected research design allows to collect data from different sources using various data collection tools. Quantitative data were collected using semi-structured and open-ended questions from key informants. Focus Group Discussion (FGD) checklists were also used to collect data. While, quantitative data was collected using questionnaire survey from the sampled fish farmers households. The questionnaire was used to collect data on the socioeconomic characteristics of the respondents, technical characteristics of the fish farms, the critical issues (water shortage, water quality, floods, and other vices), institutional framework, and respondents' perceptions of the environmental changes. Experienced male (six) and female (six) enumerators were recruited and trained.

Sample and Sample Techniques

A multistage sample technique was used to select the study area and respondents. Egypt and Nigeria were purposely selected because of the green and blue water-type availability for aquaculture, respectively. In Egypt, Kafr El Sheikh (KFS), Bahera, Shakia, and Port Said Governorates were purposely selected due to their aquaculture practice. Likewise in Nigeria, Ekiti, Ondo, Osun, and Oyo states were selected purposely in southwest of the country. This is because the states are in the same geopolitical zones. In the Egypt and Nigeria sites, a total of 160 fish farmers' households were selected using proportional random sampling method (see Table 1). The data collection was carried out between January and April, 2017, in Egypt and Nigeria.

Table 1 Distribution of questionnaire survey participants sample fish farmer households

	Egypt			Nigeria		
No.	Site	Total fish farmer households	Sample size	Site	Total fish farmer households	Sample size
1.	Kafr El Sheikh	35	35	Ekiti state	20	20
2.	Bahera	15	15	Ondo state	20	20
3.	Shakia	15	15	Osun state	20	20
4.	Port Said	15	15	Oyo state	20	20

Source: Authors Computed Field Survey

Data Collection Tools

Data collections were done by a six paired (male and female) trained enumerators who were trained on the use of Kobocollect and questionnaire administration. Data collections were done using Kobocollect application on the tablets/IPhones; in some cases, structured questionnaires, the completed questionnaire, were later keyed into the Kobocollect. Interview schedule and Focus Group Discussions (FGD) were also employed to get responses from the fish farmers in accordance to the set objectives and scope of the research work. KoboToolBar supported the participants to collect data online and offline especially during a logistical constraint. It gives room for backed up into secure servers. The questionnaire with similar content was prepared in two languages: English and Arabic. The English version was administered in Nigeria, while the Arabic version was administered on the respondents in Egypt.

Results and Discussion

Socioeconomic Characteristics of Fish Farmers in Egypt and Nigeria

The results of the Kobocollect analysis of the fish farmers' socioeconomic characteristics are presented in Table 2.

Gender: refers to the socially built characteristics such as norms, roles, and relationship of and between groups of women and men (male or female). Like age classification, it also varies from society to society. The result showed that the percentage of male fish farmers in Nigeria was 93% and that of the female was 7%. Though, this chapter revealed there was no female fish farmers in Egypt. This suggests that females might have been involved in other activities apart from fishing. The result also revealed that more males were in aquaculture than the females in both countries. This might be due to the fact that aquaculture production is more of manual labor/blue-collar job in which more physical energy are required. They are also likely to be involved in other fish farming activities like processing and marketing couple with their home cares which were sometimes uncounted (Adeleke 2013). The result also agrees with the study of Olufayo (2012); Adeleke and Fagbenro (2013) that women participate not only in the traditional fisheries sectors of fish processing and marketing but also in the nontraditional sectors of aquaculture, fisheries research, education, and extension.

Fish farming as main occupation: In Egypt, 89% of the respondents took fish farming as the main occupation, whereas the rest (11%) considered it as their secondary occupation. On the other hand, the result revealed that 41% of the respondents in Nigeria ranked fish farming as the main occupation, while majority (59%) chose it as the secondary occupation; other occupation engaged by the respondents in the study areas were civil servant, trading, carpentry, tailoring, teaching, agribusiness, and driver. This suggests that aquaculture in Nigeria is still

Table 2 Socioeconomics characteristics of the fish farmers in Egypt and Nigeria

Category or variable	Egypt Frequency (%)	Nigeria Frequency (%)
Gender		
Male	79 (100.0)	74 (92.5)
Female	–	6 (7.5)
Total	79 (100.0)	80 (100.00)
Fish farm as main occupation		
Yes	70 (88.6)	33 (41.25)
No	9 (11.4)	47 (58.75)
Total	79 (100.0)	80 (100.00)
	Mean ± S.E.	**Mean ± S.E.**
Fish farming experience (years)	18.6 ± 1.10	6.46 ± 1.22
Farm distance from residents (km)	22.27 ± 6.41	4.7 ± 1.07
Number of household	6.28 ± 0.34	4.57 ± 2.83
Number of children under 18 years	2.96 ± 0.24	1.4 ± 1.17
Number of adult above 60 years	0.19 ± 0.07	0.41 ± 0.29
Number of male in the household	3.27 ± 0.19	2.36 ± 2.46
Number of female in the household	3.00 ± 0.21	2.2 ± 1.67

Source: Authors Computed Field Survey

at the developmental stage. Aquaculture production in Nigeria at present is fairly insignificant at 5000 tons/year (FAO 2000). FAO estimates that this could be increased to about 650,000 tons. Recent researches showed that aquaculture in Nigeria has potential to contribute to both food security and economic development of the country but is still in a promising business. Despite the fact that aquaculture is an age-old practice in Egypt, it is still contributing immensely to the country's GDP, employment, and food security as it is doing relatively in most countries and is significant to the economic, livelihood, and nutritional welfare of the populace.

Technical Characteristics of Fish Farming System and Operation in Egypt and Nigeria

Parameters such as scale of operation, types of pond, source of water, access to water, water quality parameters, types of fish produced/cultured species, fish with highest production and demand, venture profitability, continuity, and inputs sources were considered. The results were analyzed and presented accordingly.

Scale of operation and purpose for rearing fish: The study revealed that small-, medium-, and large-scale farming operations were practiced in the study areas (see Table 3). The scale of operations determined the purpose of rearing fish as designated by this study. Small-scale production implies for household consumption; medium-scale production is meant for household consumptions and sales, while large-scale production does mean for fully commercial/industrial/exports. The

Table 3 Technical characteristics of fish farming system and operation in Egypt and Nigeria

Indicators	Egypt Frequency (%)	Nigeria Frequency (%)
Purpose for rearing fish		
Small scale (household)	–	50 (62.5)
Medium scale (household and sales)	1 (1.3)	24 (30.0)
Large scale (commercial/industrial/exports)	78 (98.7)	6 (7.5)
Total	79 (100)	80 (100)
Types of pond		
Earthen ponds	78 (98.7)	70 (87.5)
Others (concrete, plastic, collapsible tanks, vats)	1 (1.3)	10 (12.5)
Total	79 (100)	80 (100)
Source of water		
Stream	–	29 (36.25)
River	5 (6.3)	9 (11.25)
Groundwater/well/boreholes	–	28 (35.0)
Rain catchment/reservoirs	–	16 (20.0)
Lake water	10 (12.7)	–
Agricultural drainage water	61 (77.7)	–
Mixed drainage and Nile water	3 (3.8)	–
Total	79 (100)	80 (100)
Problems as regards the water source		
Yes	63 (79.7)	33 (41.25)
No	16 (20.3)	47 (58.75)
Total	79 (100)	80 (100)
Have you taken precautionary measures		
Yes	45 (57.0)	21 (26.25)
No	34 (43.0)	59 (73.75)
Total	79 (100)	80 (100)

Source: Authors Computed Field Survey

results revealed that none of the respondents in Egypt practice small-scale fish production; only 1% of them were operating their farms on medium scale, while 99% operate more than ten Fadden. This shows that fish production in Egypt is in a large-scale operation.

On the other hand, 63%% and 30% in Nigeria represented those that operate their fish farms on small scale and medium scale, respectively, while only 7% operated their fish farms on a large scale. The result also revealed that aquaculture in Nigeria is in small scale (93%) as compared to Egypt. The result indicates that fish farming operation in Nigeria is still at the subsistence level where production is mainly for the households with little or nothing for the market/sale. This implies that Nigeria will depend on import of fish to meet the demand of the common people.

The result supports that Egypt has achieved the scale of expansion and growth in aquaculture (Adeleke and Matthias 2017). Hence, aquaculture in Egypt is a 2.2 billion USD a year industry and has been supporting the country's GDP in recent years, making it the ninth largest global aquaculture producer in 2012 and by far the largest in Africa and the Middle East (FAO 2016). Efforts should be geared toward encouraging and promoting aquaculture practices in Nigeria in order to meet the demand of the households and consumers. Large-scale production should also be encouraged in order to expand the local and international market.

Types of pond: 99% of the fish farmers in Egypt cultured their fish in earthen ponds, while in Nigeria, 88% cultured their fish in earthen ponds. One percent and 12% of the fish farmers in Egypt and Nigeria, respectively, represent other types of ponds such as concrete, plastics, collapsible, and fiber tanks and vats. The results of the study revealed that earthen ponds are commonly used in both countries. The percentages of other types of ponds like plastics or tanks are more in Nigeria (12%) than in Egypt (1%). Because of the culture and environment, tilapia is most commonly grown fish species in Egypt. It can do well in earthen ponds where natural food is required; it cannot survive in very low or too high temperature. In contrast in Nigeria, catfish can be raised in an enclosure or tanks where it can thrive, and no artificial aeration is required, but artificial food is required for the survival of the enclosed fish. It implies that ponds type is a very vital parameter to be considered in aquaculture business.

Source of water: the results of the study revealed that fish ponds can get water from various water sources depending on the availability, location, and law. In Egypt, the following water sources and percentages were identified: lake (13%), river (6%), agricultural drainage water (78%), and mixed drainage and Nile (45%). It was observed that most of the fish farmers relied on agricultural drainage water. The result is not far-fetched from the fact that aquaculture is the last user of water in Egypt because it made used of reused water. According to the water policy in Egypt. On the other hand, the percentages of the sources of water in Nigeria are represented as follows: 36% stream, 11% river, 35% groundwater/boreholes, and 20% rain catchment/reservoirs. It was observed that in Nigeria, most of the water was fresh-water sources; this also determined the type of fish to be cultured.

Apparently Egypt represents a good example of a green aquaculture because 78% of the fish farmers depend solely on agricultural drainage water while Nigeria is good representation of a blue aquaculture because the water sources were mainly fresh water hence, the type of fish to be cultured. In Egypt, recent research showed that the use of aquaculture drainage water for agriculture is better preferred to the existing practice of agriculture drainage water for aquaculture. The use of aquaculture drainage water is best practice due to its numerous advantages. Aquaculture drainage water contained a lot of nutrients for the crops than the agricultural drainage water which might at the same time not be good for the cultured fish. Efforts should be geared to convince the government in Egypt to pay more attention to the reused of aquaculture drainage water for effective and maximum utilization of water in the country. Eighty percent (80%) of the fish farmers in Egypt that the source/s of the water to their farms/ponds is/are challenging while only 20% believe the source/s of water is/are good and not challenging (Table 3). Despite the natural and blue water sources in Nigeria, 41% believed the water source/s posed a serious problem to their fish farms, while 59% were comfortable with the water source/s to their fish farms.

In Egypt, the problem that was highlighted and emphasized mainly on water sources was pollution, while in Nigeria, the problem was water shortage as a result of dry season. In Egypt, 57% of the respondents have taken precautionary measures as regards reducing the effect of pollution on the water source that enters their farms by checking the water qualities, filter the water when pumping, treatment of water and pond, but 43% are yet to take any precautionary measures. The reason might be due to the fact that they are constrained to the available water source to be used for aquaculture in the country. In Nigeria, 60% of the fish farmers have alternative source of water such as getting water from other ponds, buying water from water tankers, damming of stream water to pump into the pond, boreholes, or well, and reducing stocking density, to their fish farms during dry season in Nigeria, while 40% of the fish farmers have not taken any precautionary measures. The reason might not be far-fetched from the facts that most of the fish farmers in Nigeria practice on subsistence level, and lack of capital is also a limiting factor from the Focus Group Discussion (FGD). Freshwater is the major water source in Nigeria and there are no restriction as to the use of water in the country, is a great opportunity and advantage for citizens who are willing to practices aquaculture to really venture into it because aquaculture is "no restriction of person."

Environmental Changes Impacts on Fish Production in Egypt and Nigeria

Table 4 presents the impacts of environmental changes on aquaculture in the study areas. The study revealed that 76% of the fish farmers in Egypt were aware of the environmental changes and its impacts on aquaculture. From the FGD, the fish farmers highlighted some observed effect of environment change on the fish production in their countries. In Egypt, the following were highlighted: reduction in rain and increase in level of temperature particularly during summer period (July); movement of the fish in the pond; sudden suffocation/floating of fish and death; low oxygen; extreme cold during winter; increase rate of fish mortality; weather change; negative change in fish feeding habit; observable differences in weather temperature; breeding time changes as the weather changes; outbreak of diseases that affect the fish resulting into fish death; sudden death of fish without reasons; change in climate in Europe; appearance of disease; increase humidity; and extreme temperature. While in Nigeria, change in raining pattern and hot temperature; pollution of air; observation of weather change; reduction in water level; negative change in fish feeding habit; flood; present of climate that is affecting hatchery; and changes in fish habit were the observed environmental changes in aquaculture practice in the country.

The research presented the evidences that there were similarities and disparities in environmental changes manifestations in both countries. The level of environmental changes' awareness of the respondents in both countries might be as a result of their level of education (Table 2). Education gave them edge on the changes that are occurring in their environment either positively or otherwise. It implies that

Table 4 Environmental changes impacts on fish production in Egypt and Nigeria

Parameters	Egypt Frequency (%)	Nigeria Frequency (%)
Do you have idea of what climate change means		
Yes	60 (76)	76 (95)
No	19 (24)	4 (05)
Total	79 (100.0)	80 (100.0)
Time to first experience the change		
Value (years)		
0–5	46 (58.3)	38 (47.5)
6–10	10 (12.7)	11 (13.8)
11–20	3 (3.8)	3 (3.8)
21–30	–	3 (3.8)
31 and above	20 (25.4)	25 (31.3)
Total	79 (100.0)	80 (100.0)
Means of environmental changes awareness		
Value (source)		
Media	29 (36.7)[a]	66 (82.5)[a]
History	64 (81.0)[a]	53 (66.3)[a]
Personal observation and fish farming experience	–	10 (12.5)[a]
Workshop/studies	–	5 (6.3)[a]
Total	79 (100.0)	80 (100.0)

Source: Authors Computed Field Survey
[a]Multiple responses

respondents in both countries are acquaintance with the changes in their environments (Maddison 2007; Nhemachena et al. 2014). Respondents' multiple responses on other means of environmental changes' awareness might not be unlikely from the global climate change program going round the world because it has brought much anxiety to all countries of the world, hence, the multiple responses.

Environmental Changes Parameters in the Study Areas

Some environmental parameters such as temperature, rainfall, wind speed, solar radiation, waves and tides, and relative humidity were further considered in the study to ascertain their effect on aquaculture practice and production in both countries. The result revealed that 99% of the fish farmers in Egypt were aware of the change in environment as a result of temperature. Sixty-three percent ascertained that rainfall pattern and volume has also changed in the country. Fifty-four percent believed that there are changes in the environment as a result of the change in wind speed. Also, 61% perceived that the change in solar radiation has affected the environment and impacted their fish production. Waves and tides are other environmental parameters that 18% of the fish farmers in Egypt perceive it change and effect on fish production. Also, 92% of the respondents further ascertained the fact that changes in relative humidity have impacted their fish production.

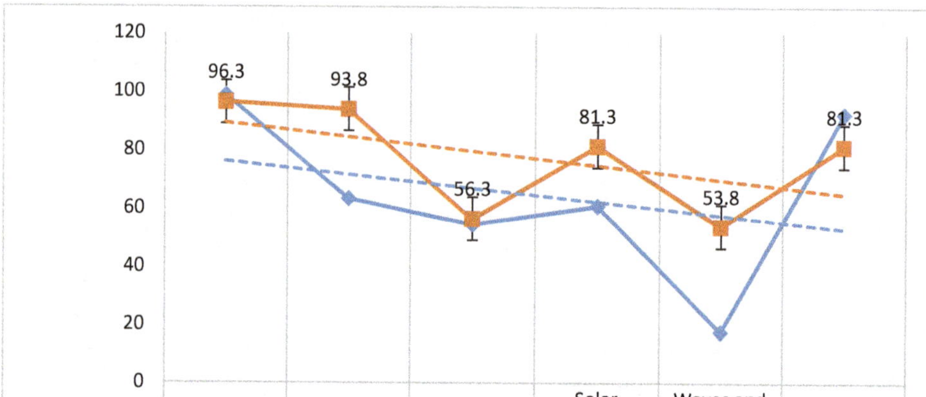

	Temperature	Rainfall	Wind speed	Solar radiation	Waves and tides	Humidity
Egypt	98.7	63.3	54.4	60.8	17.7	92.4
Nigeria	96.3	93.8	56.3	81.3	53.8	81.3

Fig. 3 Environmental changes parameters in the study areas. (Source: Authors Computed Field Survey)

In Nigeria, 96% of the respondents ascertained that temperature change is an environmental factor that affects their fish production in the country. Ninety-four percent perceived change in rainfall pattern and volume, 56% perceived change in wind speed, 81% perceived change in solar radiation, and 54% perceived change in waves and tides, while 81% also perceived change in relative humidity as environmental factors that have change in recent years (Fig. 3 and Table 4).

The FGD revealed the effects of these environmental parameters on fish production in both countries. Death of fish/mortality; reduction in fish production/growth rate; frozen of fish in winter/low temperature; rise in cost of inputs; and weakness/low appetite in fish were observed in Egypt. Some of the measures taken by the fish farmers to alleviate the adverse effect of the environmental changes were to improve their management practices and also change their production season as the environment is changing. The effects on aquaculture as expressed by the fish farmers in Nigeria were as follows: low fish production; low feeding in fish; reduction in fish growth rate; more stress for farmers; color of water changes; flooding; and death of fish. Based on the Focus Group Discussion (FGD), the following measures were adopted by the fish farmers in Nigeria change feeding formula/pattern and time, low stocking, pumping or alternative water sources, constant change of water, study/observe the weather before feeding, minimize feeding in the morning and more in the evening, stop stocking, freshwater and good feeding, removing fish to the other ponds and applying alternative means.

Critical Issues on the Study

Change in fish production in recent years in the study areas: Change here means positive or negative, increase/decrease, and adverse/mild depending on the previous position of operation of the fish farms/farmers. In Egypt, 96% confirmed that there has been change in fish production in recent years. This might have been to the various problems such as change in climatic factors which have direct impacts on fish production and pollution of water to the fish farms (agricultural drainage water). The chapter reveals that, climate change affects and impacts agricultural crops, the end users of the irrigated water will suffer directly or indirectly from the effects. Policy makers should look into the plight of the fish farmers in the country and make available use of other means of water to aquaculture. In the contest of Nigeria, the fish farmers (71%) did not affirm any serious problems during the course of this study. The reason might be to the fact that aquaculture is still at a promising stage which is best described as the first stage of production (stage 1) where there are still increasing rate of returns (IRR).

Other critical issues observed from the study as presented in Table 5. Most of the respondents (92% in Egypt and 79% in Nigeria) do not belong to any cooperative society. Bearing in mind the numerous advantages of belonging to a group, unlike in other section of agriculture, e.g., the Cocoa Farmers Association in Nigeria. In addition, the fish farmers in both countries were deprived of the benefits of the

Table 5 Critical issues

Parameters	Egypt Frequency (%)	Nigeria Frequency (%)
Has there been any change in fish production in recent years		
Yes	60 (95.9)	23 (28.75)
No	19 (24.1)	57 (71.25)
Total	79 (100.0)	80 (100.0)
Does your fish production meet the demand of the consumers		
Yes	35 (44.3)	44 (55.0)
No	44 (65.7)	36 (45.0)
Total	79 (100.0)	80 (100.0)
Major challenges faced in fish farming		
Finance	–	43 (53.75)
Labor/man power	–	14 (17.5)
Fish feed cost and other inputs	27 (34.2)	6 (7.5)
Water shortage/water quality	32 (40.5)	11 (13.75)
Fish disease	20 (25.3)	–
Vices (thief and flood)	–	6 (7.5)
Total	79 (100.0)	80 (100.0)
Do you belong to any cooperative society		
Yes	6 (7.6)	17 (21.25)
No	75 (92.4)	63 (78.75)
Total	79 (100.0)	80 (100.0)
Do you have access to extension agents		
Yes	17 (21.5)	8 (10.0)
No	62 (78.5)	72 (90.0)
Total	79 (100.0)	80 (100.0)

Source: Authors Computed Field Survey

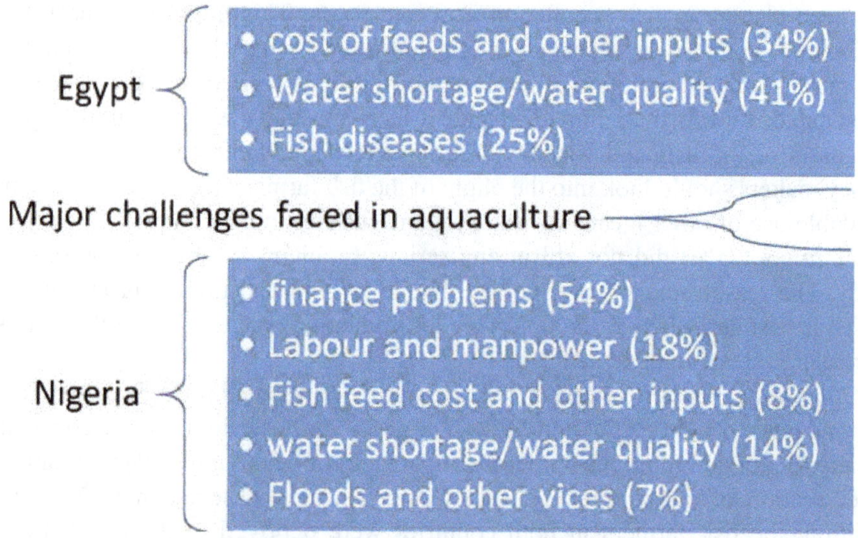

Schematic 1 Major challenges faced in aquaculture in the study areas

extension agents as revealed by the results of the study: Egypt 76% while Nigeria 90%. These issues were observed as critical; hence, urgent or future attention should be provided in order to protect the future strides of aquaculture in both countries.

Major challenges faced in aquaculture in the study areas: The cost of fish feeds and other inputs contributed to the challenges facing aquaculture; 34% of the fish famers established this fact. The result buttressed the dependence on private input suppliers who supplied most of the fish farmers on credit and later get all the gains/profit of the fish farmers; hence, the fish farmers have no say in his farm. That is why most of the fish farmers wanted to quit fish business in Egypt. Forty-one percent were facing water problems; this has to do with the agricultural drainage that they were using, and many of them were not comfortable but have no choice because it is the entitled water for aquaculture in the country. Twenty-five percent were faced with fish disease problems. This suggests urgent call for policy makers for intervention. Finance is the major problem facing aquaculture in Nigeria. Most of the respondents (54%) confirmed that finance is the major challenge unlike in Egypt where water is the major delinquent (Schematic 1).

Conclusion and Recommendations

The research revealed that there were similarities and disparities in environmental changes indications in Egypt and Nigeria. Though the level of environmental changes awareness of the respondents in both countries varies, the fish farmers were aware of the environmental changes and its impacts on aquaculture. Aquaculture is an age-old practice in Egypt; however it is still contributing immensely to the country's GDP, employment, and food security as it is doing relatively in most countries and is significant to the economic, livelihood, and nutritional welfare of the population. Conversely, Nigeria's aquaculture has the potential to contribute to both food security and economic development of the country, but it is still in a promising business. Efforts should be geared to address the critical issues such as climate change and other environmental factors that affect aquaculture practice in Nigeria. Continental and regional integration of aquaculture practices should be encouraged in Nigeria and other parts of Africa.

References

Adeleke ML (2013) Socioeconomic characteristics of the artisanal fisherfolks in the coastal region of Ondo State, Nigeria. J Econ Sustain Dev 4(2):2013133. www.iiste.org. ISSN 2222-1700 (Paper) ISSN 2222-2855 (Online).

Adeleke ML, Balogun AA (2013) Occurrence of climate change and its effects on the fishing activities in the coastal region of Ondo State, Nigeria. Int J Dev Econ Sustain 1(1):1–13

Adeleke ML, Fagbenro OA (2013) Livelihood diversification and operational techniques of the artisanal fisherfolks in the coastal region of Ondo State, Nigeria. Int J Innov Res Dev 2(1):262–273

Adeleke ML, Matthias W (2017) Chapter 13: Adaptation of the artisanal fisher folks to climate change in the coastal region of Ondo State, Nigeria. In: Leal Filho W (ed) Innovation in climate change adaptation, climate change management. Springer, Cham, pp 177–193. https://doi.org/10.1007/978-3-319-25814-0_13

FAO (2000) Fishery country profile, Nigeria. FID/CP/NIR Rev. 4. Food and Agriculture Organization of the United Nations, Rome

FAO (2010) National Aquaculture Sector Overview. Egypt. National Aquaculture Sector Overview fact sheets. Text by Salem AM, Saleh MA In: FAO Fisheries and Aquaculture Department [online], Rome. Updated 16 Nov 2010 (cited 17 Apr 2015). http://www.fao.org/fishery/countrysector/naso_egypt/en

FAO (2012) The state of world fisheries and aquaculture 2012. FAO, Rome

Food and Agricultural Organisation FAO (2016) The state of world fisheries and aquaculture 2016. Contributing to food security and nutrition for all. Rome. 200 pp

French GT, Awosika LF, Ibe CE (1995) Sea level rise and Nigeria: potential impacts and consequences. J Coast Res Spec Issue 14:224–242

GAINS Reports (2015) The state and development of aquaculture in Egypt. https://gain.fas.usda.gov/.../The%20State%20and%20Development%20of%20Aquacu.... Nov 6, 2016

Jhingran VG (1987) Lectures at ARAC for the Senior Aquaculturists course. Nigerian Institute for Oceanography and Marine Research Project RAF/82/009

Maddison D (2007) The perception of and adaptation to climate change in Africa. Policy research working paper 4308. The World Bank, Development Research Group, Sustainable Rural and Urban Development Team, Washington, DC

Nhemachena C, Mano R, Mudombi S, Muwanigwa V (2014) Perceptions on climate change and its impact on livelihoods in Hwange district, Zimbabwe. Jàmbá J Disaster Risk Stud 6(1):123. https://doi.org/10.4102/jamba.v6i1.123

Olufayo MO (2012) The gender roles of women in aquaculture and food security in nigeria. IIFET 2012 Tanzania Proceedings

Sumaila UR, Christophe B, Alice T (2014) Fishing for the future: trends and issues in global fisheries trade. December 2014

Worldometers (2017) Elaboration of data by United Nations, Department of Economic and Social Affairs, Population Division. World population prospects: 2017. www.worldometers.info

World Bank (2013) The Fish to 2030: Prospect for Fisheries and Aquaculture World Bank Report Number 83177-GLB (December 2013)

10

Do-It-Yourself Flood Risk Adaptation Strategies in the Neighborhoods of Kano City, Nigeria

Aliyu Barau and Aliyu Sani Wada

Contents

Abstract

The urban poor in developing countries is hit hardest by climate-related extreme events such as flooding. Also, informal settlements lacking municipal support and immediate public response to flooding incur losses and thus exacerbate their sufferings. Left out or left alone, the vulnerable people from some parts of the ancient city of Kano develop their own efforts to protect themselves against the recurrent flood events. Hence, this chapter examines the nature of community-driven do-it-yourself (DIY) adaptation The data was collected through field-based surveys, interviews, and questionnaires to enable in-depth analysis of the problem from socioecological point of view. The results identified flood drivers to include

A. Barau (✉) · A. S. Wada
Department of Urban and Regional Planning, Bayero University Kano, Kano, Nigeria
e-mail: asbarau.urp@buk.edu.ng; saniwada2@gmail.com

the nature of surface topography, torrential rainfalls, lapses, and inadequacies in the availability of drainage infrastructure and human behavioral lapses in drainage management. On the other hand, the DIY adaptation manifests in the use of sandbags, de-siltation of drainage, construction of fences, and drainage diversions. It is important to highlight that DIY adaptation is a good strategy; however, municipal authorities must come to the aid of such communities and revisit the absence of urban planning by supporting them through capacity building to find more effective solutions to the challenges of the changing climate and environment.

Keywords

Planning · Households · Climate Change · Vulnerability · Informal

Introduction

Climate-related disasters continue to spread and affect the poorer populations in both urban and rural areas. It has been suggested that flooding is the most widespread climate extreme event that accounts for more than 40% of disasters globally (Nkwunonwo et al. 2016). It is also considered as the third most damaging disaster after storms and earthquakes (Wilby and Keenan 2012). It was estimated that the average annual number of people affected by floods is likely to increase from one million in the 1990s to 25 million by 2050 (Salami et al. 2016). Nevertheless, the increased awareness and concerns over climate change have reinforced the need for adjustments on how human use land and other resources in order to limit risks and vulnerabilities that originate from global climate change (Verburg et al. 2012). Besides climate change, many studies have attributed the exacerbation of flooding to the nature of population density and standard of living especially in the lived environments of developing countries (Devkota et al. 2014). In other words, this implies that built environments in developing countries are most exposed to flooding and hence more response actions are needed from within and outside the lived environment. Efforts from within involve action by individuals and municipalities, and action from outside may involve states and federal actions to strengthen resilience to flooding and climate change. Importantly, it is critical to also put in place modalities for communicating climate knowledge to public in order to enable proper response to climate crisis (Barau and Tanko 2018).

Researchers from developing and developed countries consider flood adaptation from different approaches. For instance, Alfieri et al. (2016) focused on adaptation measures for flood risk reduction by putting emphasis on adaptive measures such as relocation, reduction of vulnerability, reduction of peak flows, and increased flood protection levels. On the other hand, Barau (2008) focused on adaptation measures oriented toward institutional perception but fails to assess public adaptation strategies deployed to prevent flooding. Some authors suggested that relief materials including healthcare services supplied to flood victims in Nigeria were inefficient and paid too little attention given to efforts to reduce the future flood menace (Soneye 2014; Olanrewaju et al. 2019). Contrastingly, since the 1990s, the United States introduced the Community Rating System (CRS), a community-driven program that incentivizes local actions for addressing flood risk (Noonan and Sadiq 2018). It is apparent that climate change adaptation is not an exclusive business of the state or formal planning processes, but it goes down to efforts of communities and businesses to protect themselves (Colenbrander and Barau 2019; Udelsmann Rodrigues 2019). By implication, climate adaptation has become a concern of every stakeholder given the weakening resilience of urban areas in developing countries. Nevertheless, designing neighborhood or community resilience to flooding will benefit from understanding the social and ecological attributes of societies which are not limited to biophysical changes, infrastructure, mitigation planning, and public perceptions of extreme events (Parsons 2019). The potentials of grassroots action to address flooding in African countries are said to be a promising and effective self-defense mechanism in poorly serviced areas (Udelsmann Rodrigues 2019).

Flooding in Nigeria is the most common and recurring disaster with unprecedented damages caused by heavy rainfall and surface runoff which in 2012 in particular contributed to submergence of settlements along the Kainji, Jebba, and Shiroro dams in Niger State, the Lagdo dam in Cameroun located along the River Benue, the Kiri Dam on the River Gongola, and numerous additional irrigation dams (PDNA 2013). Consequently, this triggered the displacement of thousands of people and loss of lives and properties. According to the National Emergency Management Agency (NEMA), 363 people were killed, 5,851 injured, 3,891,314 affected, and 3, 871, 53 displaced due to the resulting floods. Similarly, it is has been observed that flood disaster happens to be more frequent in Kano, Niger, Adamawa, Oyo, and Jigawa states possibly due to influence of river Benue, Niger and Ogun (Nkwunonwo et al. 2016).

Since the 1960s, the rainfall received in Kano region has exhibited notable fluctuations, however with increased flood incidents due to encroachment on environmentally sensitive areas such as floodplains along the ancient city walls which were traditionally reserved specifically for urban gardening (Barau et al. 2015). Studies revealed that between 2003 and 2007, Kano has experienced severe urban flood that claimed many lives and property – roughly 16, 730 households were affected, 14 people were reported dead – and many were displaced (Barau 2008). In addition, flood water contaminates and pollutes the environment with garbage; many factors are responsible for the recurrent flood events with prolong rainfall leading to large volume of water with tendency of overwhelming the infiltration capacity of the soil. This often yield to large surface runoff beyond the carrying capacity of the already narrowed drainages filled with garbage and simultaneously adjoined with poor maintenance which end up polluting the environment (Azua et al. 2019). It is apparent that damages to lives and property has become a cyclic phenomenon in Kano metropolis. It was also reported that 18, 397, 29, 151, and 177 people in 2003, 2004, 2005, 2006, and 2007, respectively, were affected in the Dala Local Government Area, due to impact of flood (Barau 2008). The aim of this chapter is to assess flood risk and adaptation measures in Dala Local Government Area in Kano city, Nigeria, and this is with a view to improving adaptation strategies in the area by identifying suitable and sustainable planning recommendations.

An Overview of Conceptualization of Flood Risk Adaptation

Floods can occur within a matter of minutes, and its effects can be local, impacting a neighborhood or a community; it can even go beyond a single community to affect the entire river basins and multiple areas (Wilby and Keenan 2012). Flood risk and flood-related losses may increase due to human activity or decrease through an appropriate flood management, planning, and adaptation strategies (Kubal et al. 2009). Human interventions in the natural processes such as increase in settlement areas, removal of biotic habitats, population growth, and economic assets over low-lying plains make them more prone to flooding especially when alterations of natural drainage and river basin patterns and deforestation are on the rise (Muo 2015). This situation is typical of what obtains from urban areas of developing countries including Nigeria which has the largest population in Africa. Hence, flooding and its impact on population is a socioecological challenge as it has links and feedbacks between human and natural systems. The socioecological system approach is said to be important in private flood hazard adaptation strategies (Fuchs et al. 2017).

Cirella and Iyalomhe (2019) categorized flooding on the basis of spatiality which includes coastal flooding and river flooding, while Dhiman et al. (2019) added urban and areal flooding to the list. In a situation whereby towns and cities are located on these types of landscapes and in the case where the most vulnerable populations are located on such marginal lands, the struggle with risks of sea-level rise becomes even more complicated. Another factor is the accumulation of rainwater in low-lying areas with a high water table, or inadequate storm drainage (Cirella and Iyalomhe 2019). In addition to this, the main causes of flooding as posited by same author are not limited to heavy rainfall, blockage of waterways, building on flood plains, and poor land use planning. According to Dhiman (2019), urbanization was the most important driver of flood risk in the cities where inadequate drainage system in combination with ineffective spatial planning increases the flood risk. Additional damages from the floods are also included in the framework where simulation using climate model and further downscaling indicate high potential sensitivity of flood risk to climate change.

Flooding with its numerous and varied impact causes mortalities, population displacement, and property damage which may also critically endanger sustainable development pathways (Vojtek and Vojteková 2016). The literature on flood health impact rarely quantifies or classifies the causes and circumstances of deaths which in many instances are also determined by the prevailing socioeconomic and health conditions of the affected communities (Jongman 2018). Flood impacts range from socioeconomic, psychological, cultural, to environmental impacts (Alfieri et al. 2016). The economic impacts include damage to public buildings, public utility works, housing and household assets, losses in industry, and business trade. Other losses are those affecting petty shopkeepers, low-income earners, loss of revenue, and interruption of road and railway transportation. Flooding also triggers the outbreak of infectious diseases, e.g., fever, pneumonic plagues, dysentery, and common cold among others.

On the other hand, Devkota et al. (2014) identified the following responses of adaptation strategies at community level undertaken before, during, and after flood event. Responses identified before flood event include management plans, keeping contact information, pre-estimating of flood risk, and producing human resources/ trained manpower. Response strategies undertaken during the flood event at community level include perfect communication at community level and selection of makeshift settlement. Proper assessment of flood risk in urban areas is still challenging simply because of its complexity as compared to situation in rural areas. Common flood risk analyses do not usually incorporate social and ecological impacts but exclusively assess economic damages, which can be measured in monetary terms (Haque et al. 2012). Comprehensive approaches which try to integrate economic, ecological, and social impacts are less available (Schanze 2006). Particularly, the latter are frequently omitted due to a lack of suitable data.

There are many studies with different conceptual frameworks for flood vulnerability, risk assessment, and adaptation. For instance, Oculi and Stephenson (2018) observed that one of the earliest "frameworks" of adaptation were based on three attributes: climatic-stimuli, the system that is adapting, and method of adaptation. They also recognized that adaptation strategies can be grouped by timeframe of interest, types of behavior, sector, scale, and level of governance. Adaptation to extreme events such as flood is vital in the present era due to the multifaceted nature of the causes and effects, which is exacerbated due to the increasing population and climate change. The ever-changing and cross-boundary nature of flood risk renders adaptation strategies as very crucial in this period of fluctuating social and environmental conditions (Jongman 2018). Flood adaptation strategies are often intended to minimize impacts on various sectors such as built environment, human health, water quality, and transport infrastructures (Wilby and Keenan 2012). The efficient and effective adaptation strategies should have the services and facilities to slow flood waters; use nature-based solutions, early warning systems, and financing schemes to tackle and reduce the economic loss; and practice risk-informed land planning (Jongman 2018).

Observing the Nature of Flood Risk Situation in Kano City

The sites observed in this chapter are located in Dala Local Government Area (LGA), which is the nucleus of the ancient Kano city in northern Nigeria. It is named after the Dala, a residual hill which is the original site from which the city emerged since prehistoric times. The area is located around between latitude 12° 00′ 00″ N to 12° 03′ 21″ N and Longitude 8° 27′ 30″ E to 8° 31′ 40″ E. (Fig. 1).

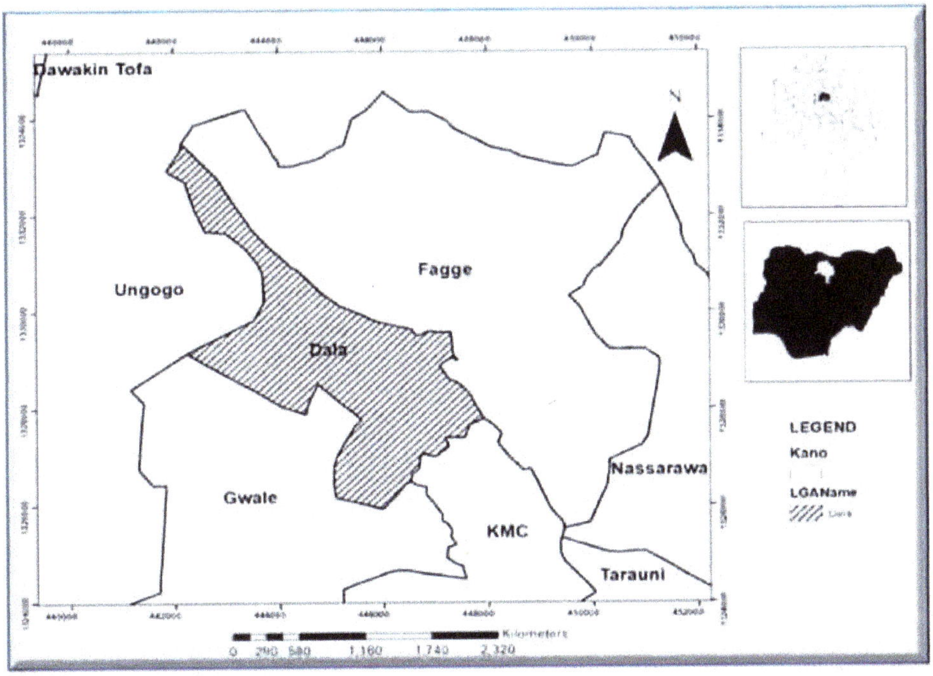

Fig. 1 Map of urban Kano showing the location of Dala Local Government Area

This LGA covers a total land area of about 19 km^2 and total population of 415,777 as of 2006 national population and housing census. There was significant increase in the natural population in the state and the study area. The natural increase was estimated to have risen from 2.51% in 1960s to 3.3% in 1980s and 4.5% till date, although the urban growth rate is about 5.5% annually (Mustapha et al. 2014). This implies that the area is a high density core urban settlement and given its nature of terrain is vulnerable to flooding. Previous studies suggested the importance of socioecological systems in understanding the urban environment in Kano (Barau et al. 2013).

Relief and Geology

Relief of the state is said to be 80% of composed basement which comprises of marbles, granites, and gneisses (Salami et al. 2016). Jakara River is underlain by crystalline Basement complex of pre-Cambrian origin which losses its identity by disappearing into the Chad Formation. Crystalline basement of pre-Cambrian forms the underline of River Jakara, made up of granites extending to Gabasawa toward East and Yadai toward North. This river serves as the natural drain in the study which collects runoff from different tributaries. This geological setting suggests that the study area is susceptible to flood based on its largely impervious surface.

Climate and Hydrology

The climate of the state is tropical wet and dry type according to Koppen's classification, although climatic changes are believed to have occurred in the past. The climate features in Kano as a whole are typical of West African savannah. Temperature in the region is generally high all year round, together with gradual increase every season. It can be said to be higher usually from January to April which may reach as high as 43 °C, respectively (Mustapha et al. 2014). According to Sanni and Okhimamhe (2009), the annual rainfall values are unpredictable from 1 year to another. For instance, 1999, 2004, 2010, and 2012 witnessed highest annual values in both temperature and rainfall, that is to say, above normal, compared to previous years.

Research Data

Primary data were collected using questionnaire, field observations, field surveys, and in-depth interview. The questionnaires were administered to household heads and/or adult respondents in each residence that were within the 100 m radius of flood-prone areas. This strategy enabled the researchers to identify factors that exacerbate flooding and types of response undertaken to lessen flood risk. Initially, the authors worked with communities to identify the flood hotspots in addition to observations made during previous flood events in the study area.

Global Positioning System (GPS) was used during field survey to take the coordinates of the flood-prone areas which was later imported into the Google Earth for spatial representation of the flood prone areas which was also used in generating the Digital Elevation Model (DEM) and contour map of the study area to analyze vividly the nature of terrain as well as contour of the study area. ArcGIS 10.4 was used to generate the location map and flood-prone areas respectively. Google Earth was used to produce the satellite imagery which was used for spatial representation, while interview was used to source institutional responses to flood risk in the study area, respectively. Interview was conducted in State Emergency Relief and Rehabilitation Agency (SERERA) and Kano State Urban Planning and Development Authority (KNUPDA) with key informant on flood risk and adaptation measures in the study area. This further aid in determining the extent to which these institutions respond to flooding.

Secondary data was sourced through searching the keywords with the aid of Google Scholar (i.e., flood, risk, flood risk, adaptation, and adaptation measures) from scholarly peer-reviewed journals, books, and government archive which provided proper understanding of the research problem at local and global level. The information sourced provided an insight to the researcher on flood risk, flood impacts, its causes, as well as factors that exacerbate flooding.

The questionnaire was administered to the seven (7) flood-prone clusters (462 houses) which include Kofar Ruwa, Kwanar Taya, Gwammaja, Babban Layi, behind Filin Dantata, Lunkwi, and adjacent Gidan Baban Gwari, respectively. Two hundred and seventeen (217) copies of questionnaires were randomly distributed to the residents within 100 meter radius of each flood-prone area (31 for each zone for equal and unbiased representation) using the simple random sampling techniques. However, only one hundred and ninety six (196) copies of the distributed questionnaires were retrieved due to loss and failure to fill by the residents. Questions asked were tailored toward identification of causes of flood, factors that exacerbate flooding, as well as resident adaptation measures to flood risk in the study area. In addition, data related to demography and socioeconomic characteristics were also captured. The sampling size adopted for this study was determined using Taro Yamane's formula (1967) with 95% confidence level. The formula is given as

$$\mathrm{n} = \frac{N}{1 + N(e)2}$$

where

$n =$ Sample size required
$N =$ Number of population in which the sample will be drawn (462 houses)
$e =$ Margin of error, given as 0.05
$1 =$ Constant.

$$\mathrm{n} = \frac{462}{1 + 462(0.05)^2}$$

$n = 214.3851508121/7 = 30.626450116$. (Akukwe 2014)

After calculating the sample size by substituting the number of dwelling units (462) into the Yamane formula, the number of sample was 214.3851508121 houses. Dividing this number by the number of flood-prone areas (7) yielded 30.626450116 houses respectively. Therefore, in order to obtain a reliable data, researcher rounded up the sample size to 31 dwellings per each flood-prone area (Fig. 2). Thus, a total of 217 sample size emerged out of 462 sample frame. For convenience and transparency, nine (9) variables were used which include heavy rainfall, poor drainage facilities, slope and topography of the area, dumping of refuse in drainages, blockage of drainage channels, impervious surface, absence of drainage channels, and unplanned development along floodplains. The targeted respondents on a normal circumstances were the household heads, but in a scenario where the household heads were not around, an adult respondent from each dwelling was selected to fill in the required questions based on 5-point Likert scale from Strongly Agree (SA = 5) to Strongly Disagree (SD = 1). This procedure was adopted from a study conducted by Devkota et al. (2014).

Institutional responses were obtained through an in-depth interview with key informants from Kano State Emergency Relief and Rehabilitation Agency (SERERA) and Kano State Urban Planning and Development Authority (KNUPDA). In SERERA, the interview was oriented toward past flood damages,

Fig. 2 Google image showing flood-prone areas and houses within the 100 m radius of each other. (Source: Authors' Field Survey 2019)

existing adaptive measures to respond to flood hazard for pre-, during, and post-flood events. This information was relevant to the study because it helped in determining the extent to which the stipulated hazard in the study area was responded to. In KNUPDA, issues raised were concerning planning regulations around the flood-prone areas. This information was relevant to the study because it further helped the researchers to bring out the level of compliance (public) to adapt to the problems of flood-prone area or otherwise. In all the cases, the interview was recorded (with permission of respondents) using phone audio recorder, which was subsequently transcribed to a verbatim record.

Five-point Likert scale and weighted average index (WAI) on the basis of "Highly Applicable" to "Very Less Applicable" were used for analysis of the questionnaire. The first part of the questionnaire was demographic and socioeconomic information of the respondents. These were analyzed using SPSS, and the results were represented in simple tables, charts, and graphs. ArcGIS 10.4 and Surfer software were used to draw and analyze the digital elevation model and contour map of the study area. While community responses were collected before, during, and after flooding by the public were analyzed using weighted average index (WAI) against a scale of 1 to 5. The adaptation strategies for each respondent were rank based on the weighted average index (WAI) which is given as

$$\mathrm{WAI} = \frac{\mathrm{F1} \times \mathrm{W1} + \mathrm{F2} \times \mathrm{W2} + \mathrm{F3} \times \mathrm{W3} + \mathrm{F4} \times \mathrm{W4} + \mathrm{F5} \times \mathrm{W5}}{\mathrm{F}_1 + \mathrm{F}_2 + \mathrm{F}_3 + \mathrm{F}_4 + \mathrm{F}_5}$$

$$\mathrm{WAI} = \frac{\sum fi \times \mathrm{W}i}{\sum \mathrm{fi}}$$

where

F = frequency of the respondents
W = weight of each scale
i = weight

Note (5 = highly applicable, 4 = applicable, 3 = moderately applicable, 2 = less applicable and 1 = very less applicable) (Devkota et al. 2014). This generated the measures and/or strategies that were mostly applicable and least applicable by the respondents in the study area.

Digital Elevation Model (DEM) of the Study Area

The digital elevation model (DEM) as illustrated in Fig. 3 revealed the elevation of the study area from which the blue shades represent the areas with the highest elevation 520–525 m above sea level (Dala hill), while the green represents the areas with the lowest elevation 470–475 in the study area respectively. These green shades on the DEM represent the locations of the flood-prone areas. It is apparent that after rainfall the stormwater radiates in a centrifugal direction reaching the lowest elevation (flood prone areas) as natural drains. However, as built-up areas, they easily become flooded.

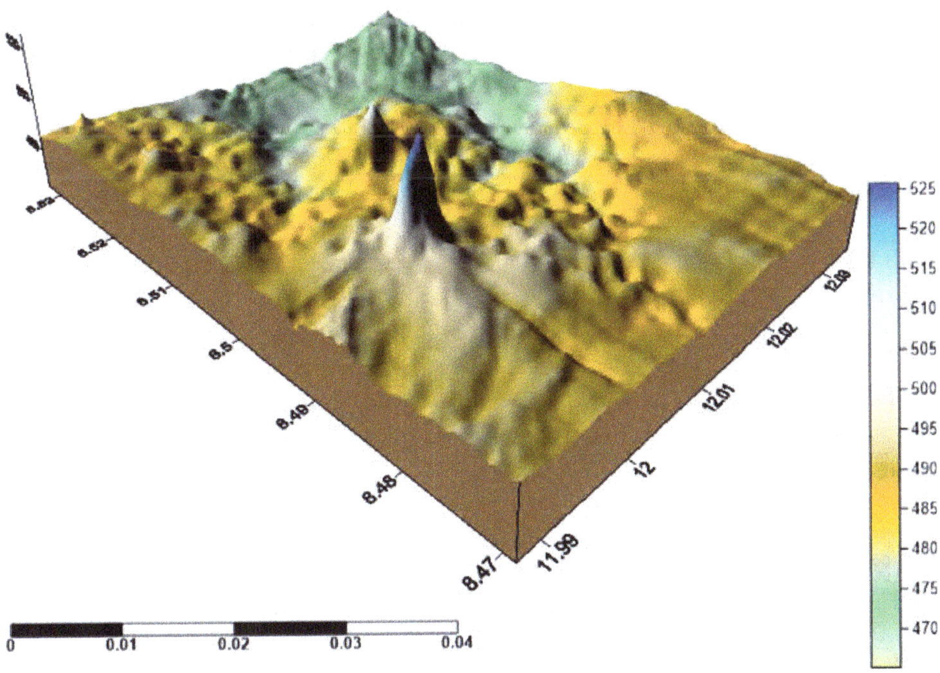

Fig. 3 Digital elevation model (DEM) of the study area showing variation of elevation

Fig. 4 Contour map of the study area. The lighter spots depict areas with highland, while the darker spots represent areas with low land

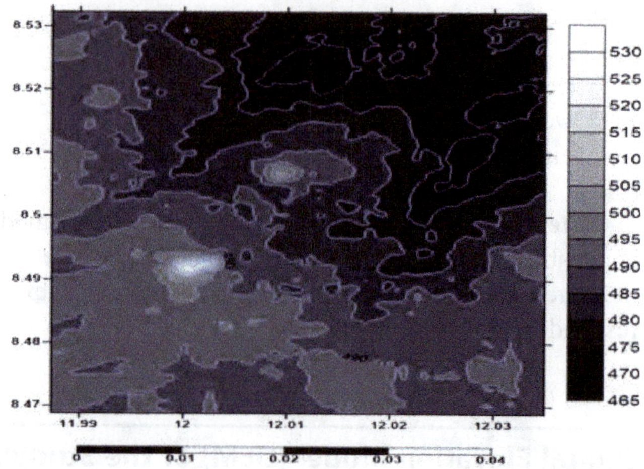

The green shades represent lowlands, while the blue shades represent highland areas (Dala hill) in the study area.

On the other hand, the contour map (Fig. 4) illustrates the variation of elevation of the study area. The white portion represents the point with the highest elevation (530 m), while the dark portion represents the points with lowest elevation (465–470 m). From these, it can be deduced that elevation still determines floodability of even highly urbanized areas.

Socioeconomic Characteristics of the Residents

The tendency of flood risk is not the same for all people although all may be situated in a flood-prone areas. It varies considerably with demographic and social characteristics. Although flood risk presents significant health burden worldwide, socioeconomic status is closely related to being flooded or otherwise (Aerts et al. 2018). This section explains the socioeconomic characteristics of the residents in the study area which cover residence, gender, age, marital status, educational level, occupation, average monthly income, housing condition, and construction material used.

Socioeconomic Characteristics of the Respondent in the Study Area

Table 4 revealed the respondent residential locations of the respondents out of which (28) 14.3% were from Kofar Ruwa, (26) 13.3% from Kwanar Taya, (30) 15.3% from Gwammaja, (27) 13.8% from Babban Layi, (29)14.8% from behind Filin Dantata, (30) 15.3% from Lunkwi, and (26) 13.3% from Gidan Baban Gwari which constitute the sum of 196 respondents. Gender is another important variable in any research of this nature. The gender distribution of respondents revealed dominance of male respondents. Male respondents featured more than female as they dominate household headship. Again their female counterparts were not freely available to interact with researchers due to cultural barriers in the study area. In general, there is an indication that majority of the respondents are still in their active and productive age. Age is an important attribute that helps in understanding people's perception about a given phenomenon. And with regard to flood risk and adaptation, it helps in improving the methods to be used in adapting to risk as well as mitigating its impact.

The perception and attitudes of an individual due to marital status may differ; findings presented in Table 4 showed the dominance of married respondents in the study area. In addition, marital status is relevant to the study because married people will be more willing to initiate coping strategies in trying to reduce potential losses than the unmarried population simply because they are more attached to the community, and it is assumed that each married person has apartment which is in the long run liable to flood risk. Household size is important to understand socioeconomic status and vulnerability of households. In this context, it is assumed that the larger the household size, the more is the risk of losing more lives and/or property and vice versa.

Knowledge influences reasoning, attitudes, and understanding of individuals from a given social phenomenon. The findings presented in Table 4 also outlined respondents' level of education where (71) 36.2% has tertiary education and (53) 27.0% with informal education, while (59) 30.1% and (13) 6.6% were having only secondary and primary level of education, respectively. This is an indication that majority of the respondents were literate and thus capable of providing adequate and relevant information with regard to flood risk and adaptation measures in the study area.

Occupation or livelihood is another important factor of consideration as it determines people's choice of residence. Thus, the social class of individuals determines their area of residence. It has been established that the more comfortable and secured areas of residence are occupied by the richer households while the less privileged are left with the marginal areas (Salami et al. 2016). In addition, income is an important determinant which is congruent with income and occupation. The situation in the study area in relation to income further highlights what is known with regard to the poor who occupy areas associated with unplanned urban growth and substandard informal settlements on floodplains among other things (Salami et al. 2016).

Age of building is another important determinants of consideration simply because building has a lifespan just like humans. Buildings deteriorate with time and thus become more exposed to destructive flood impacts. Findings revealed that with exclusion of some buildings among which are Government Girls College, Dala, and National Orthopaedic Hospital majority of the buildings in the area have undergone transformation. This implies that majority of the buildings were constructed not so long, thus a concrete evidence of recent encroachment on open spaces and changes in building materials. Building materials play important role in proofing water absorption. Such changes in urban morphology are said to have increased flood events in urban Kano which covers the study area (Barau 2015) (Table 1).

Table 1 Showing frequency and percentage response on the socioeconomic characteristics of the respondent

Gender	Frequency	Percent
Male	150	76.5
Female	46	23.5
Total	**196**	**100.0**
Age		
18–25 years	57	29.1
26–32 years	51	26.0
33–39 years	54	27.6
40 years and above	34	17.3
Total.	**196**	**100.0**
Marital status		
Single	90	45.9
Married	106	54.1
Total	**196**	**100.0**
Household size		
0–4	82	41.8
5–9	52	26.5
10–14	49	25.0
15 and above...	13	6.6
Total	**196**	**100.0**
Occupation		
Civil servants	32	16.3
Traders	97	49.5
Farmers	21	10.7
Others	46	23.5
Total	**196**	**100.0**
Average monthly income		
<20,000	63	32.1
20,000–40,000	49	25.0
40,000–60,000	36	18.4
60,000 and above	21	10.7
Total	**196**	**100.0**
Age of building		
0–5 years	46	23.5
5–10 years	78	39.8
10–15 years	41	20.9
15 years and above	31	15.8
Total	**196**	**100.0**
Building construction material		
Mud	40	20.4
Cement block	154	78.6
Bamboo	0	0
Others	2	1

(continued)

Table 1 (continued)

Gender	Frequency	Percent
Total	**196**	**100.0**
Condition of buildings		
Need minor repair	101	51.5
Need major repair	39	19.9
In good condition	56	28.6
Total	**196**	**100.0**

Public Perception and Current State of Flood Risk and Adaptation

Adaptation may include modifying susceptibility, increasing response capacity, and reducing exposure (Merz et al. 2010); thus, this section focused on community perception of flood risk, vulnerability profiles of the households regarding flood experience, as well as the residents' adaptation (coping) mechanisms against flood risk (i.e., actions taken by the public prior to, during, and after flood event) in the study area. Questionnaire survey was used to assess the level of flood severity and whether or not affected previously. The following parameters "severely affected," "affected but not severe," and/or "not affected at all" were used in generating this information which revealed that (68) 34.7% of the respondents were not affected at all, (91) 46.4% were affected but not severe, while (37) 18.9% of the respondents were severely affected by previous flood havoc in the study area. In order words, it can be deduced that 65.3% of the sampled population were affected by flood disaster in the study area (Fig. 5).

The factors that cause and exacerbate flooding in the study area based on the outcome of the current study include heavy rainfall, dumping of refuse on drainages, as well as blockage of drainage among others as shown in Table 2. These factors were ranked based on average mean score (X) of each factor and sum weighted value (SWV) of responses derived from questionnaires.

Findings as presented in Table 2 established that the indicator with the highest rating is rainfall with a weighted value of 4.77 and a positive deviation of 0.82 from the mean, while the least rated indicator was impervious surfaces with a weighted value of 3.08 and a negative deviation of −0.87 from the mean, respectively. This denotes that heavy rainfall is the major cause of flood in the study area. Thus, climate change appears to be a major factor and driver of flooding incidents in urban areas of developing countries. Therefore, it is imperative for urban areas to embrace priority action to tackle climate change both as local and global challenge (Bai et al. 2018). The study further revealed that out of the nine indicators used, five had a positive deviation which are the major causes of flooding in the study area, while four had a negative deviation from the mean. The indicators with positive deviation about the mean include heavy rainfall (with a weighted value of 4.77 and a positive mean of 0.82), dumping of refuse and/or waste on drainages (with a weighted value of 4.42 and a positive mean of

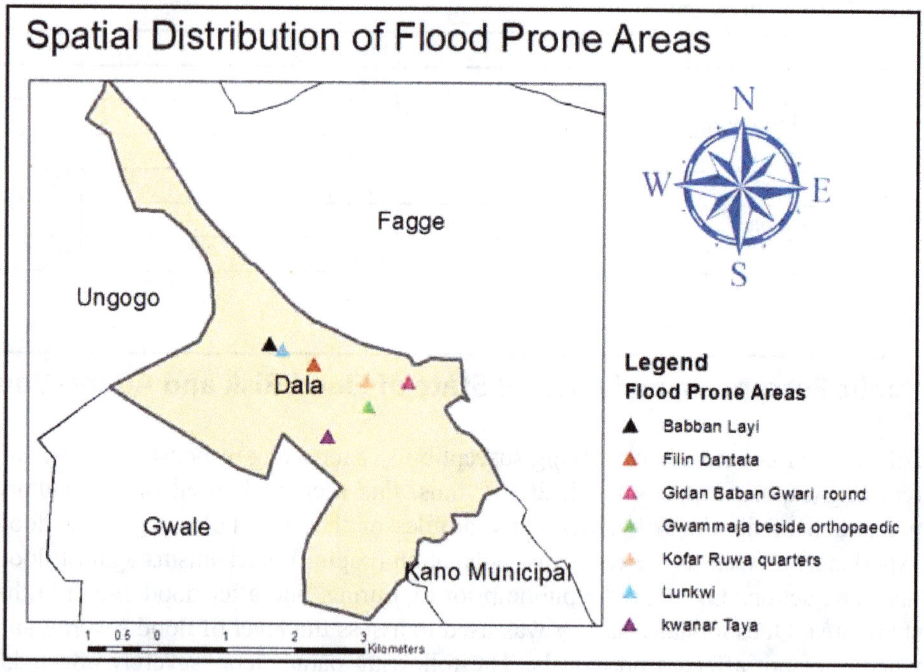

Fig. 5 Spatial distribution of the flood-prone areas in Dala LGA. (Source: Author 2019)

Table 2 Causes of flood in the study area with their ranks in descending order

Causes	SA	A	U	D	SD	SWV	X	X̄ -X	Ranks
Heavy rainfall	154	40	0	2	0	934	4.77	0.82	1st
dumping of waste/ refuse on drainages	110	70	4	12	0	866	4.42	0.47	2nd
Slope and topography of the area	86	66	22	20	2	802	4.09	0.14	3rd
Blockage of drainage channels	82	77	9	21	7	794	4.05	0.1	4th
Poor drainage facilities	79	69	26	22	0	793	4.05	0.1	5th
Unplanned development along flood plain	40	97	41	15	3	744	3.80	−0.15	6th
Absence of drainage channels	63	64	28	37	4	733	3.74	−0.21	7th
Soil type	11	72	68	38	7	630	3.21	−0.37	8th
Impervious surface	26	41	77	26	26	603	3.08	−0.87	9th
Total	**651**	**596**	**275**	**193**	**49**	**6973**	**35.58**	–	–

Source: Authors' Field Survey

Note: **SA** (5) = STRONGLY AGREE, **A** (4) = AGREE, **U** (3) = UNDECIDED, **D** (2) = DISAGREE, **SD** (1) = STRONGLY DISAGREE, **SWV** = Sum of Weighted Value, **X** = Mean score. **X̄** = Mean deviation

0.47), slope and topography of the area (with a weighted value of 4.09 and a positive mean of 0.14), blockage of drainage channels (with a weighted value of 4.05 and a positive mean of 0.1), and poor drainage facilities (with a weighted value of 4.05 and a positive mean of 0.1). On the other hand, the indicators with negative deviation about the mean were unplanned development along floodplains (with a weighted value of 3.80 and a negative mean of −0.15), absence of drainage channels (with a weighted value of 3.74 and a negative mean of −0.21), soil type (with a weighted value of 3.21 and a negative mean of −0.37), and impervious surface (with a weighted value of 3.08 and a negative mean of −0.87).

The results presented here highlight the importance of socioecological dimensions of flooding because of close and tangential relations between human and naturally induced drivers of climate change, landscape, ecology, and infrastructural development. It is equally important to recognize that the impact of human transformation of landscape and/or misuse of infrastructure galvanize negative effects of climate change in the society. For instance, as evident from Figs. 6, 7, and 8, people erect structure in areas where the morphology is not favorable, and hence flood waters cause destruction to buildings (Fig. 6). On the other hand, indiscriminate dumping of waste causes siltation and sedimentation of waterways which cause flooding in the study area (Figs. 7, 8, and 9). In other words, people need to be educated and informed about climate change as it is obvious that level of education does not sufficiently make people more informed. Hence, it is critical to follow multiple strategies of educating and communicating climate change to the public (Barau and Tanko 2018).

Fig. 6 Abandoned development in a low-lying area due to adverse effect of flooding in the study area (Kofar Ruwa quarters). (Source: Field Survey 2019)

Fig. 7 Poor drainage filled up with silt, obstructing the free flow of storm water along Dala Orthopaedics Gwammaja. (Source: Field Survey 2019)

Fig. 8 Indiscriminate dumping of waste along drainage in the study area. (Source: Field Survey 2019)

Public and Institutional Adaptation Response to Floods in the Study Area

Public and institutional response to flood risk is one of the objectives of the study. Here, public response to flood risk was collected through observation and questionnaire survey, while the institutional response was collected through an interview with key informants in the specified planning and environment agencies. Firstly, with

Fig. 9 Area submerged during the 2019 flood in Kwanar Taya. (Source authors' field work 2019)

regard to people's effort to prevent flood risks which some researchers called it do-it-yourself (DIY) adaptation (Udelsmann Rodrigues 2019), a field survey revealed that the residents of the study area also employed this strategy through the use of sand bags, raising of walls, and de-siltation among others to divert flood water away from their residential area (community), and this was specifically observed in Gwammaja besides Dala National Orthopedics Hospital and also at Kwanar Taya which is another section of the study site. All these efforts represent the community's local responses to reduce potential harm and adverse effects of flood incidents. However, due to uncertainty of rainfall and anthropogenic factors that induce the floods, the community members can hardly achieve complete defense of their locality as it is beyond their capacity to prevent floods. This situation remarkably improve our understanding of the DIY adaptation efforts and at the same time the indispensability of municipalities to support the efforts of vulnerable communities particularly those inhabiting marginal areas that are more exposed to climate change.

The study further revealed that most of these areas that were subjected to frequent flood hazard have developed different flood risk management approaches for pre-flood, during, and post-flood period. Table 3 outlines the residents' coping strategies undertaken before flood/rainy season. Clearing of waterways appeared to be the first with WAI of 4.45 and followed by de-siltation as the second most practiced strategy with WAI of 3.77; widening of drainage channels was ranked third with WAI of 3.51, and producing human resource/trained manpower was ranked fourth with WAI of 3.48. Likewise use of metal sheeting and/or sandbag was ranked fifth with WAI of 3.26, while sharing of contact information was ranked sixth with WAI of 2.86 as the least strategy practiced in the study area, respectively.

Table 3 Pre-flood adaptation strategies at community level (N = 196)

Response	HA	A	MA	LA	VLA	SMV	WAI ($\bar{X}$)	X	($\bar{X}$-X)	Ranks
Clear water ways to remove dirt	128	41	21	0	6	873	4.45	3.56	0.89	1st
De-siltation	38	115	21	3	19	738	3.77		0.21	2nd
Share contact information	15	53	41	63	24	560	2.86		−0.7	6th
Produce trained manpower	44	60	56	19	17	638	3.48		−0.08	4th
Widened drainage channels	26	106	27	16	21	688	3.51		−0.05	3rd
Use of metal sheeting/ sandbags	25	84	37	17	33	639	3.26		−0.3	5th
Total	**276**	**459**	**203**	**118**	**304**	**4136**	**21.33**			

Source: Author's Field Survey 2019
Note: **HA** = Highly Applicable, **A** = Applicable, **MA** = Moderately Applicable, **LA** = Less Applicable, **VLA** = Very Less Applicable, **WAI** = weighted Average Index, **SMV** = sum of mean variable

The current study has further revealed that community adaptation planning is undertaken at various scales reflecting, in particular, the seasonality of rainfall. For instance, among the 196 respondents, 128 (63.5%) stated that clearing of drainages in anticipation of rainy season was a highly considered strategy for improving the drainage capacity and its functionality. This strategy was regarded as the most relevant flood risk adaptation strategy in the study area which is followed by 115 (58.7%) of the respondent that mentioned de-siltation as the next most important adaptation strategy that is used in the community (see Table 3). Again, it is important to note that local governments have actually failed in delivering their responsibility of environmental upkeep which can be carried out through increased financing and enforcement environmental and sanitary provisions. In other words, while the DIY role played by the communities is commendable, the total neglect of vulnerable communities is deplorable. In the interest of environmental justice and social equality and inclusion, it becomes imperative on local governments in developing countries to consider integrated and larger adaptation strategy that will tackle community level vulnerabilities and strategic efforts such as DIY as a means of achieving an inclusive adaptation to flood risk. The need to focus on flooding is very important as it has been identified as universal issue in global climate change risk which affects more people than any other form of disaster (Nkwunonwo et al. 2016; Salami et al. 2016).

Flood Adaptation Strategies at Community Level

The above subsection dwelt on pre-flood DIY efforts, while Table 4 in this subsection shows the residents' coping strategies during flood/rainy season based on applicability of various adaptation strategies. According to the outcome of the survey, selection of safer place to stay was ranked first (most applicable) in the study area with 3.85 WAI. Whereas relocation to the nearest neighborhood until after rainfall was second with 3.82 WAI, the third most applicable strategy was taking care of the children, the aged, and people living with disability (PWDs) with 3.68 WAI. Here, the results point out the critical role of decision-making as to how to respond to rainy and floodable days in these urban communities. One of the limitations of this study is its inability to explore the nature of decision-making with respect to displacement of people.

Moreover, the study revealed that out of the total respondents, 86 (43.9%) respondents mentioned selecting safer location to stay, while 71 (36.2%) mentioned taking care of the children, the aged, and PWDs as the most applicable strategy practiced during flood in the community. Similarly, temporary relocation to nearest neighborhood until after flooding 112 (57.1%), perfect communication at community level for assistance 65 (33.2%), and taking care of the affected people 73 (37.2%) were enlisted among applicable (although not highly applicable) strategies in the community for adapting to flood risk during flood. The relocation sites are mostly homes of friends and relations that have raised buildings and more spaces to accommodate people. Even under this scenario, DIY adaptation strategy is very active and effective. For instance, digging of channels to divert waters 70 (35.7%) and praying for lesser rain 58 (29.6%) were mentioned among moderately applicable strategies to adapt during flood. The total loss of confidence in the government manifested clearly as 27.6% of the respondents mentioned the government intervention.

Post-flood Community Response on Adaptation Strategies

Flooding can have immediate-, short-, and long-term impacts on urban communities. As such, community adaptation responses may extend to beyond the rainy season. In the context of this study, communities in the study area lack the appropriate solutions but hang on the adaptation strategies that they have developed over time. Nevertheless, our questionnaire created a kind of public conversation among the respondents who revealed their thoughts on what seems to be their perceptions of best adaptation strategies. In this regard, Table 5 outlines the roles of clearing blocked drainages in the community which was ranked as the most applicable flood adaptation strategies with a WAI of 4.11. People themselves have acknowledged the role of education as a driver of community adaptation, and this knowledge will in a long way support and enhance their capacity for DIY adaptation. This further revealed that no relief materials were received by the public in these flood-prone areas. Again, as can be seen in Table 5, the respondents stressed the need for equality in distribution of relief

Table 4 During flood adaptation strategies at community level (N = 196)

S/N	Response	HA	A	MA	LA	VLA	SMV	WAI ($\bar{X}$)	X	($\bar{X}$-X)	Ranks
1	Select safer location to stay	86	64	6	11	29	755	3.85	3.41	0.44	1st
2	Temporary relocate to the nearest neighborhood until after flooding	40	112	22	12	10	748	3.82		0.41	2nd
3	Take care of the children, the aged and PWD's	71	56	31	11	27	721	3.68		0.27	3rd
4	Perfect communication at community level for assistance	57	65	26	28	20	699	3.57		0.16	4th
5	Take care of the affected people	30	73	55	12	26	657	3.35		−0.06	5th
6	Digging channels to draw water away from dwellings	38	44	70	24	20	644	3.29		−0.12	6th
7	Pray for lesser rain	35	47	58	24	32	617	3.15		−0.26	7th
8	Do nothing and wait for government assistance	13	41	47	41	54	506	2.58		−0.83	8th
	Total	**370**	**502**	**315**	**163**	**218**	**5347**	**27.29**			

Source: Authors' Field Survey 2019

Table 5 Post-flood adaptation strategies at community level responses (N = 196)

S/N	Response	HA	A	MA	LA	VLA	SMV	WAI	X	($\bar{X}$-X)	Rank
								($\bar{X}$)			
1	Seeking of help for clearing blocked drainages in the community	120	23	28	4	21	805	4.11		1.43	1st
2	Inform the community/ward head on areas mostly affected	15	101	41	29	10	670	3.42		0.74	2nd
3	Engage on public awareness campaign on flood risk management	40	32	40	25	59	557	2.84	2.68	0.16	3rd
4	Collaborate with Government and other agencies	0	2	23	56	115	304	1.55		−1.13	4th
5	Equal distribution of relief resources, if any	0	11	23	17	145	292	1.49		−1.19	5th
	Total	**175**	**169**	**155**	**131**	**350**	**2628**	**13.41**			

Source: Authors field work 2019

Fig. 10 Resident adaptation measures (coping strategy) to flood, to hold back and/or divert flood water away from their residence (Gwammaja)

materials from the public sector if and when at least they come to the victims of flooding. In other words, there is seemingly lack of trust on the government and its agencies for their failure to assist people. While it has been a culture for governments at all levels (local, state, federal) to allocate for disasters such as the recurrent floods, there is dissatisfaction from the side of the vulnerable communities. Again, as can be seen in Figs. 10 and 11, there is evidence of community DIY actions that aim at protecting neighborhoods and their households from recurrent flood incidents.

Institutional Flood Adaptation Response Strategies in the Study Area

Public institutions such as agencies have duty and responsibility to promote adaptation measures within their area of constitutional jurisdiction and governance system operational in a given urban area. Against this backdrop, this study sought for explanations from relevant agencies which in this case are two organizations, namely, Kano State Urban Planning and Development Authority (KNUPDA) and State Emergency Relief and Rehabilitation Agency (SERERA). In the case of KNUPDA, our questions were answered by the Zonal Coordinator who is in charge of development control in the study area. By virtue of his knowledge of the area, he affirmed that "the problem of flood in the local government is not a new phenomenon," adding that the study site is "among the ancient settlement in the state which existed for long prior to initiation of the Control Department (CD) or KNUPDA itself." This view impliedly suggests that probably the planning authority has been

Fig. 11 De-siltation of drainages to improve free flow of storm water, but not properly evacuated which have tendency of finding its way back into the drainage

overwhelmed with the flood incident in the study area and possibly lacks sustainable solutions for the challenge. Given the experience of the United States, it is important for authorities in developing countries to tap on community initiatives in dealing with flood hazards (Noonan and Sadiq 2018).

Furthermore, the officer argued that "increased population and urbanization make the community continue to become more congested." In addition, he mentioned that there were numerous buildings that were developed without planning approval and that majority of the area was not properly planned or designed. Accordingly, such haphazard developments are among the key drivers of perennial floods. The situation is also not unconnected with inadequate drainages and encroachment of open spaces by private buildings among others. However, the officer maintained that some areas were but still suffer from lack of proper implementation of design principles and enforcement of planning regulations. Hence, he argued that "adaptation strategy employed is insisting on people to seek proper approval and development permit before construction." In doing so, KNUPDA provides prospective developers with adequate standards in developing their properties. According to him, they also convey messages to the ward heads as a mean of ensuring control of development and stop any development that encroaches floodplain in the Local Government Area which eventually might pose a problem such as environmental hazard.

With regard to the implementation and enforcement of planning regulations, the key informant asserts that "some officials are always in constant inspection of the area and these officers are responsible for reporting (immediately) any suspicious development." Another notable problem noted by the officer is the lack of proper waste collection and disposal infrastructure in the ancient city of Kano. Although poor waste disposal is invariably connected to flooding, the officer noted that waste management is no longer the business of KNUPDA as it has been assigned to another authority. Nevertheless, he insisted that KNUPDA had severally issued stop work notice and in some instances demolished hazardous structures when they are deemed to constitute a threat to safety. Some areas within and around the study area hardly experience flood incident and attributed that to their interventions in the local government area.

In SERERA, an agency responsible for responding to all forms of disasters in the state, the key informant revealed that they "don't have pre-flood adaptation measures" because they only embark on emergency response with immediate effect when it is reported to them. This agency does provide parties affected with relief materials ranging from building materials like cement, blocks, nails, and planks, among others. They also embark on assessment reports, making recommendations for relief measures, occasional involvement in the distribution of relief materials, and evacuation of stranded persons in spots of emergency. The respondents added that their personnel were trained on how to rescue individuals from either fire or flood disaster which is sent on special assignment when the need arises.

When asked on whether they have any special plans for the identified flood-prone areas under study, they responded that that they just act in case of emergency and provide relief materials; things provided include mosquito nets, foodstuffs, drugs, mattresses, temporary camping of stranded people, as well as taking care of the aged and children. Their response is mainly on immediate report of accidents, and they work hand in hand with fire service department whose mandate include rescue services.

Based on information given by the key informants, it is obvious there is disconnect between the approaches and philosophies of communities and public institutions responsible for tackling climate disasters and, in particular, flooding. The recurrent floods cause loss and damage to communities, and this situation increases temporary loss of shelter or displacement from their houses. This is an indication of the situation in many developing countries where there is disconnect between planning policy and climate change mitigation and adaptation. It is obvious there is need for capacity building and reorientation of urban climate change governance architectures to enable creation of a broad-based consultations and policy formulations between communities and public agencies. This is necessary because at the moment actions by communities and the agencies do not complement each other. At the same time, there is fragmentation even between and within public organizations responsible for urban planning, waste management, and rescue operations.

Conclusion

This chapter is premised on the socioecological notions which enables to gain a better understanding of flood adaptation response in urban areas of developing countries. In other words, it is imperative to look into the connections of human and social dimension of environmental and climate change as we observe climate vulnerability in cities and towns. This chapter has established the critical role of the physical environmental features such as relief and topography and their interface with flood incidence in high-density urban areas. Thus, it is crucial to recognize the role of physical terrain when planning and designing human settlements in this age of changing climate. In relation to this, municipalities in developing countries should prioritize the importance of systematic environmental data in urban planning climate adaptation strategies. The recurrence of flooding in cities and towns cannot be disconnected with neglect of terrain factor in stormwater overflows and its redistribution within the built-up areas. Again, as urban surface is becoming more impervious and waterproofed, it is important to reconsider the place of nature in tackling climate change.

The neglect of nature in urban development is closely followed by neglect of people in the effective management of climate crisis and, in particular, perennial floods that ravage cities and towns. It appears that in developing countries public agencies operate in silos even when there is evidence of commitments to addressing climate challenges. Therefore, it is high time to address such gaps and disconnects which cause failures and loss of time and resources due to such unyielding interventions. Invariably, the persistence of fragmented efforts leaves public to suffer resounding devastation of infrastructure as well as people's dwelling places and sources of livelihoods.

More people inhabit urban areas more than ever before in human history, and this human dominance is being challenged by climate change and its impacts. The neglect of climate vulnerable groups inhabiting informal areas has continued in most developing countries. This situation violates the principles and goals of the United Nations' Sustainable Development Goals (SDGs) and, in particular, the SDG 11 – which focus on sustainable human cities and communities. In a situation like this, one of the best options left for the people left out in the schemes of policy and planning is DIY adaptation. This adaptation strategy is found to be effective and useful to many poor and neglected urban citizens in developing countries. Hence, it is imperative on local authorities to support such efforts by working with communities that use this kind of adaptation measures. What is most attractive about DIY adaptation is that it focuses on directly correcting or tackling flood causative and/or multiplier factors. Therefore, researchers and especially climate experts and scientists need to recognize DIY as part of the grassroots efforts and local knowledge and actions which are essential in building global climate adaptation.

References

Aerts JCJH, Botzen WJ, Clarke KC, Cutter SL, Hall JW, Merz B, Mysiak J, Surminski S, Kunreuther H (2018) Integrating human behaviour dynamics into flood disaster risk assessment. Nat Clim Change. https://doi.org/10.1038/s41558-018-0085-1

Akukwe TI (2014) Determinants of flooding in Port Harcourt Metropolis, Nigeria. Int J Humanit Soc Sci 19(11):64–72

Alfieri L, Feyen L, Di Baldassarre G (2016) Increasing flood risk under climate change: a pan-European assessment of the benefits of four adaptation strategies. Clim Change 136(3–4): 507–521. https://doi.org/10.1007/s10584-016-1641-1

Azua S et al (2019) Spatial multi-criteria analysis for mapping of flood vulnerable areas in Fagge Local Government Area of Kano State, Nigeria. FUTY J Environ 13(1):23–35

Bai X, Dawson RJ, Ürge-Vorsatz D, Delgado GC, Salisu Barau A, Dhakal S, Dodman D, Leonardsen L, Masson-Delmotte V, Roberts DC, Schultz S (2018) Six research priorities for cities and climate change. Nature 555(7694):23–25. https://doi.org/10.1038/d41586-018-02409-z

Barau AS (2008) An appraisal of climate change risks and institutional adaptation strategies in Kano State. In Association of Nigerian Geographers 50 th Annual Conference Book of Proceedings. The Department of Geography and Regional Planning, University of Calabar, Index Book Publishers, Calabar, p 113–120

Barau A, Tanko AI (2018) In search of new narratives for informed decisions on climate change crisis in the African Drylands. In: Leal Filho W, et al (eds) Handbook of climate change communication: vol 3, Climate change management. Springer International Publishing, pp 1–20. https://doi.org/10.1007/978-3-319-70479-1_1

Barau AS, Ludin ANM, Said I (2013) Socio-ecological systems and biodiversity conservation in African city: insights from Kano Emir's Palace gardens. Urban Ecosyst. https://doi.org/10.1007/s11252-012-0276-x

Barau AS, Maconachie R, Ludin ANM, Abdulhamid A (2015) Urban morphology dynamics and environmental change in Kano, Nigeria. Land Use Policy 42:307–317

Cirella GT, Iyalomhe FO (2019) Determinants of flooding and strategies for mitigation: two-year case study of Benin City, pp 1–15. https://doi.org/10.3390/geosciences9030136

Colenbrander S, Barau A (2019) Planning and financing urban development in the context of the climate crisis. Int J Urban Sustain Dev 11(3):237–244. https://doi.org/10.1080/19463138.2019.1673529

Devkota RP, Cockfield G, Maraseni TN (2014) Perceived community-based flood adaptation strategies under climate change in Nepal. August. https://doi.org/10.1504/IJGW.2014.058758

Dhiman R, Vishnuradhan R, Arun TIE (2019) Flood risk and adaptation in Indian coastal cities: recent scenarios. Appl Water Sci 9(1):1–16. https://doi.org/10.1007/s13201-018-0881-9

Fuchs S, Karagiorgos K, Kitikidou K, Maris F, Paparrizos S, Thaler T (2017) Flood risk perception and adaptation capacity: a contribution to the socio-hydrology debate. Hydrol Earth Syst Sci. https://doi.org/10.5194/hess-21-3183-2017

Haque AN, Grafakos S, Huijsman M (2012) Participatory integrated assessment of flood protection measures for climate adaptation in Dhaka. 24(1):197–213. https://doi.org/10.1177/0956247811433538

Jongman B (2018) Effective adaptation to rising flood risk. 9–11. https://doi.org/10.1038/s41467-018-04396-1

Kubal C, Haase D, Meyer V, Scheuer S (2009) Integrated urban flood risk assessment – adapting a multicriteria approach to a city. Nat Haz Earth Syst Sci 9:1881–1895

Merz B, Hall J, Disse M, Schumann A (2010) Fluvial flood risk management in a changing world. Hydrol Earth Syst Sci 10:509–527

Muo A (2015) Evaluating the causes of flooding in six communities in Awka Anambra State of Nigeria. J Nat Sci Res *5*(4):1–10

Mustapha A, Yakudima II, Alhaji M, Nabegu AB, Adamu F, Dakata G, Umar YA, Musa BU (2014) Overview of the physical and human setting of Kano. Research Journali's Journal of Geography 1(5):1–12

Nkwunonwo UC, Whitworth M, Baily B (2016) Review article: a review and critical analysis of the efforts towards urban flood risk management in the Lagos region of Nigeria. 349–369. https://doi.org/10.5194/nhess-16-349-2016

Noonan DS, Sadiq AAA (2018) Flood risk management: exploring the impacts of the community rating system program on poverty and income inequality. Risk Anal. https://doi.org/10.1111/risa.12853

Oculi N, Stephenson SR (2018) Conceptualizing climate vulnerability: understanding the negotiating strategies of small island developing states. Environ Sci Policy 85:72–80. https://doi.org/10.1016/j.envsci.2018.03.025

Olanrewaju CC, Chitakira M, Olanrewaju OA, Louw E (2019) Impacts of flood disasters in Nigeria: A critical evaluation of health implications and management. Jamba 11(1):557

Parsons M (2019) Extreme floods and river values: a social–ecological perspective. River Res Appl. https://doi.org/10.1002/rra.3355

PDNA (2013) Nigeria Post-Disaster Needs Assessment (PDNA) 2012 Floods. May

Salami RO, Giggins H, Salami R, Meding V (2016) Vulnerability of human settlements to flood risk in the core area of Ibadan metropolis, Nigeria. Jamba: Journal of Disaster Risk Studies 9(1)1–14.

Sanni D, Okhimamhe A (2009) Linking climate change and urbanization in Kano Metropolis and environs in the Sudano-Sahelian Belt of Nigeria, IOP Conf. Series: Earth and Environmental Science 6(2009) 342032. https://doi.org/10.1088/1755-1307/6/4/342032

Schanze J (2006) Flood risk management - a basic framework. In: Schanze J, Zeman E, Marsalek J, editors. Flood risk management - hazards, vulnerability and mitigation measures. NATO Science Series. Houten: Springer Netherlands; p. 120

Soneye A (2014) An overview of humanitarian relief supply chains for victims of perennial flood disasters in Lagos, Nigeria (2010–2012). J Humanit Log Supply Chain Manag 4:179–197

Udelsmann Rodrigues C (2019) Climate change and DIY urbanism in Luanda and Maputo: new urban strategies? Int J Urban Sustain Dev. https://doi.org/10.1080/19463138.2019.1585859

Verburg PH, Koomen E, Lesschen JP (2012) An assessment of the impact of climate adaptation measures to reduce flood risk on ecosystem services. 473–486. https://doi.org/10.1007/s10980-012-9715-6

Vojtek M, Vojteková J (2016) Flood hazard and flood risk assessment at the local spatial scale: a case study. 5705. https://doi.org/10.1080/19475705.2016.1166874

Wilby RL, Keenan R (2012) Adapting to flood risk under climate change. https://doi.org/10.1177/0309133312438908

Yamane T (1967) Statistics, An Introductory Analysis, 2nd Ed., New York. Harper and Row

11

Climate Variability on Fishing Activities in Inland Waters: Case of Owena River in Ondo and Osun States, Nigeria

Lydia Adeleke, Jacob Victor Jerry, Desalegn Ayal, Akinola Joshua Oluwatobi, Ayodele Idowu Sunday and Ajibefun Igbekele Amos

Contents

L. Adeleke (✉)
Department of Fisheries and Aquaculture Technology, Federal University of Technology, Akure, Nigeria
e-mail: mladeleke@futa.edu.ng

J. V. Jerry · A. J. Oluwatobi · A. I. Sunday
Department of Fisheries and Aquaculture Technology, Federal University of Technology, Akure (FUTA), Akure, Nigeria
e-mail: jacobvictor402@yahoo.com; joesheathakinola@gmail.com; ayodeleidowusunday@gmail.com

D. Ayal
Center for Food Security Studies, College of Development Studies, Addis Ababa University, Addis Ababa, Ethiopia
e-mail: desalegn.yayeh@aau.edu.et

A. I. Amos
Department of Agricultural and Resource Economics, Federal University of Technology, Akure (FUTA), Akure, Nigeria
e-mail: iaajibefun@futa.edu.ng

Abstract

This chapter examined the effects of climate variability on fishing activities in inland waters: the case of Owena River in Ondo and Osun States, Nigeria. The particular inland water body (Owena) was selected purposively because the Owena River crosses across the two States (Ondo and Osun States). A total of 100 respondents/fisher folks were selected randomly. Primary data was used to get information from the respondents with the aid of the structured questionnaire. Secondary data was used to get information on the climate variability existing in the locations in order to achieve the set objectives. Specifically, the effects of climatic variables such as temperature, rainfall, and wind were determined on fisher folk, fishing duration, fish catch, fishing techniques, and fishing equipment/gears. Descriptive survey design was used to examine the demographic characteristics of respondents. The result revealed that most of the respondents (96%) were male. This chapter established that climate variability particularly strong wind, rainfall, and temperature has adverse effects on fishing activities such as reduction in fish catch, long duration of fishing, loss of fishing gear, change in fishing techniques, and effect on health; hence, the livelihoods of the residence are adversely affected on the long run. In addition to the adverse effect of poverty and loss of life for the fisher folks. Although, the fisher folks livelihoods depend mainly on fisheries resources and optimum fishing depends on favorable climate/weather conditions. Therefore, proper preventive coping strategies against the adverse effect of climate variables should be paramount in both States (Ondo and Ekiti) to improve livelihoods and food security.

Keywords

Climate variability · Fishing activities · Inland waters · Livelihoods · Nigeria

Introduction

Climate variability is the observable change in atmospheric conditions of a given area at a particular time usually 10 years. The Intergovernmental Panel on Climate Change defined climate change or variability as any change in climate overtime, whether due to natural variability or as a result of human activities. Temperature and precipitation are the most common climate variables critical to measure with regard to food systems. Climate variability poses a threat to the effective productivity and sustainability of fishing activities and resources around the world. Climate variabilities are changes or variation in climate condition due to factors such as anthropogenic factors, e.g., bush burning, release of fossil fuel, and other biochemical activities which will consequently lead to global warming, loss of abundant fisheries resources and other agricultural resources, and also depletion of ozone layer (Cheung et al. 2009).

Climate plays an important role in the world. Because most livelihood and living resources depend on suitable climatic condition for survival and continuity, its effects can be felt directly or indirectly. The Intergovernmental Panel on Climate Change (IPCC) predicted that during the next decades, billions of people, especially those in developing countries, will face changes in rainfall patterns that will contribute to severe water shortages or flooding, rising temperatures that will cause shifts in crop growing season, and aquatic organism distribution.

Climate change and global warming will definitely cause adverse effects on the livelihoods of the populace in Nigeria if not prevented, and it will affect crop production, livestock production, fisheries, forestry, and post-harvest activities, because the rainfall regimes and patterns will be altered, floods which devastate farmlands would occur, increase in temperature and humidity which increases pest and disease would occur, and other natural disasters like floods, ocean, and storm surges, which not only damage Nigerians' livelihood but also cause harm to life and property, would occur.

Nigeria is a West African country found in the tropics; it has only two seasons which are the dry season and the raining season. The country is one of the most populated countries in the world with over 180 million people and is blessed with abundant aquatic resources. However, Nigeria is vulnerable to climate change impacts due to its geography, location, climate, vegetation, soils, economic structure, population and settlement, energy demands, and agricultural activities.

The prominence and persistence of climate change has made it a global menace with adverse effects borne by developing nations (Agbebi and Omoniyi 2011). Global disasters perpetrated by storm-surge, flood, coastal erosion, and other natural hazards have brought much anxiety and bewilderment to the world governing agencies. These impacts and implications are profound as they occur through the increasing changes in water temperature/sea levels (Adeleke and Omoboyeje 2016).

Fisheries is one of the fastest growing sectors in the world, and it provides employment, food, income, and recreation for many people globally. The dependence of the human populace on fishing activities as a major source of food and livelihood is linked to the rich nutritional composition of fisheries products (fish, crayfish, prawn, shrimps, etc.), palatability, digestibility, and high source of animal protein, omega fatty acid, and vitamins.

Despite these numerous and enviable benefits, fisheries resources, functions, and structures have come under intense threat due to the extensive and increasing anthropogenic activities that exist in and around various water bodies. These result in a wide array of problems for both the aquatic resources users and decision makers. These anthropogenic activities often lead to induced climate change which have shown to have apparent effect on aquatic ecosystems, the economic and social system that depends on them (World Fish Centre 2006; Perry et al. 2009).

Recent studies on the effect of climate change on aquaculture production and management in Akure metropolis, Ondo State, and also the occurrences of climate change and its effect on the fishing activities in the coastal region of Ondo State (Adeleke and Omoboyeje 2016; Adeleke and Balogun 2013), respectively, have revealed the occurrences of climate variability and its effect on fishing activities through alteration in temperature, wind pattern, rainfall pattern, water qualities, and availability.

Severe effects such as great reduction in aquatic resources, low fish catch, reduced fish production, slow growth rate, increase in boat mishap, loss of lives, poverty, etc. arise mainly from increased flooding, variation in rainfall pattern and intensity, and rising temperature. Idowu et al. (2011) reported that the impacts of climate change are expected to exacerbate the impacts of human pressure which will further diminish the ability of natural ecosystems to continue to provide ecosystem services and may cause invasion of strange species that are favored by climate change thereby threatening biodiversity. Also, poor adaptation strategies and high dependence on fisheries have been reported to expose small-scale fisher folks in most developing country to climate change (Allison et al. 2009). These effects impact on fish species composition, production, and yield, risk of health and life of fisher folk, and loss and damage to livelihood assets. Therefore, the fisher folk will have to seek adaptation and mitigation strategies in order to sustain their livelihood.

The inhabitants of Ondo and Osun are mostly small-scale commercial fish farmers that depend on fisheries resources as a source of food and livelihood. However, climate variability may threaten their livelihood and infrastructural development. Hence, there is the need to identify the pathways through which climate variability and change are impacting or likely to impact the fishing-dependent communities. This will also help policy makers in implementing adaptive strategies to be able to withstand and cope with these adverse climate variability. This research work will examine the effect of climate variability on fishing activities in Owena Ondo and Osun States.

The following research questions were answered in this chapter. What are the effects in Owena, Nigeria, are of temperature, rainfall, and wind variability on fish catch; rainfall, temperature, and wind variability on fishing techniques and equipment; rainfall, temperature, and wind variability on fishing duration; and rainfall, temperature, and wind variability on fisher folks.

The Chapter Design

The chapter design that is adopted in this chapter is descriptive survey designed to obtain precise information on the current status of climate change and to possibly draw a valid conclusion from the survey. It should be noted that there are no climate change nor meteorological records in the area as observed limitation of the chapter.

Owena Ondo State and Owena Osun State were used. Owena Ondo is located at geographical coordinate 7° 12′ 0″ North, 5° 1′ 0″ East, while Owena Osun is located at latitude 7.19908 and longitude 5.018267 degree. Both areas were purposively

Fig. 1 Map of Nigeria. (Source: Authors' Download: Google map, 2018)

selected based on general fishing activities, trading, accessibility, and anthropogenic activities (Figs. 1 and 2).

Chapter Design

A multistage sampling technique was used to select the samples in this chapter. The first stage involved the purposive selection of Owena Ijesa Ori Ade Osun State and Owena Market road Ondo East Local Government and Owena Ondo State. The justification of the selection is based on the fact that the two locations are close and prominent for fishing activities and have the same water body flow to both locations. The target population are fisher folk, and fishermen in Owena Ondo State and Owena Osun State were purposively selected based on their closeness of the site and the prominent fishing activities in the area. Primary data from fisher folks were collected with the aid of an interview schedule using structured questionnaire. 50 questionnaires each were administered to both locations making it 100 in total. The interview schedules were divided into sections based on the objectives of the research. Data obtained from various sources were processed before analysis. The data were edited, coded, and entered for analysis in the Statistical Package for Social Sciences (SPSS). Summary tables were then prepared to facilitate recording and further analysis. Data obtained through questionnaire were analyzed using quantitative methods. The data

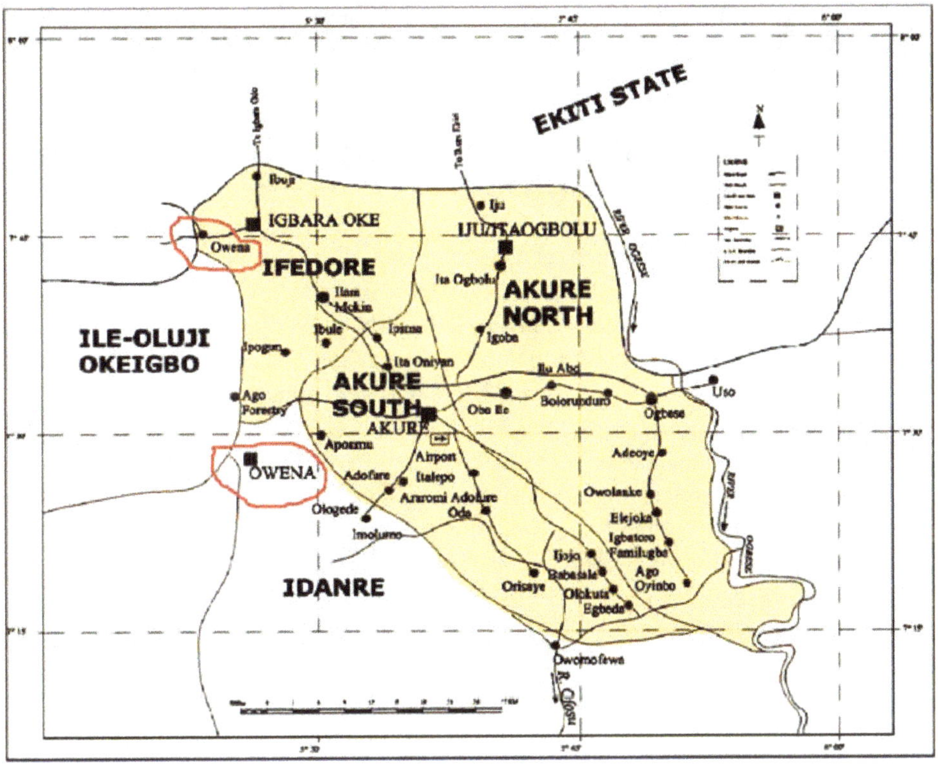

Fig. 2 Map of study area. (Source: Authors' Download: Google map, 2018)

from the questionnaire were subjected to computation of simple statistics such as frequencies, percentages, tabulations, and cross-tabulations and presented in form of tables, graphs, and charts.

A multistage sampling technique was used to select the samples in this chapter. The first stage involved the purposive selection of Owena Ijesa Ori Ade Osun State and Owena Market road Ondo East local Government.

Cronbach Alpha

Cronbach alpha (α) was used to analyze the sufficiency of the survey instrument.

$$\text{Cronbach } s\alpha = \frac{N - \overline{C}}{\overline{V} + (N-1).\overline{C}}$$

Where:

$N =$ No of items (12)
$\overline{C} =$ Average inter – item covariance among the items
$\overline{V} =$ Average variance

The survey instrument for this chapter is sufficient because Cronbach alpha is greater than 0.6.

Distribution According to Gender and Age Group of Respondent in Owena Osun and Ondo States

The distribution according to gender and age group of the populace involve in fishing activities in this chapter is presented in Figs. 3 and 4, respectively. The result revealed that all of the respondent (100%) in Owena Ondo were males, while 92% of the respondents in Owena Osun were male and 8% females. The pooled result revealed that 96% of the respondents were males while 4% were females. The high percentage of men involvement in fishing as revealed by the most of the respondents in the areas attest to high-risk associated fishing. In addition, the extensive involvement of women in post-harvest fish processing and fish trading might have accounted for their low percentage in fishing in both States inland water environs.

The mainstream of the respondent (52%) were between 21and 31 years, 38% of the respondents were between 31and 40 years, 8% <21 years, and 2% were between 41 and 50 years. This implies that the pooled age brackets were within the age group of 21–45 years. This age bracket is considered to be the most economical, vibrant, resourceful, and productive group when considering the energy required to carry out a tedious activity like fishing and the need to source for income and livelihood.

Secondary Occupation and Monthly Income of Respondent in Owena Osun and Ondo State

The secondary occupation and monthly income of the populace involve in fishing activities in the areas is presented in Figs. 5 and 6, respectively.

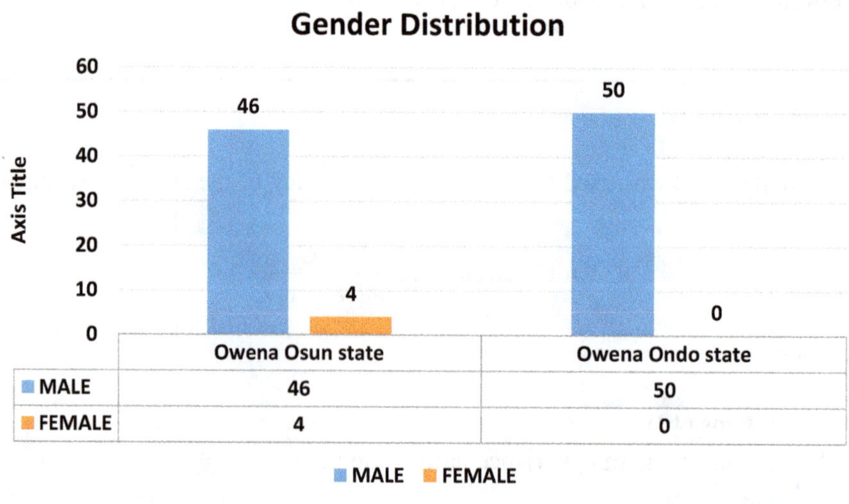

	Owena Osun state	Owena Ondo state
MALE	46	50
FEMALE	4	0

Fig. 3 Gender distribution of respondents

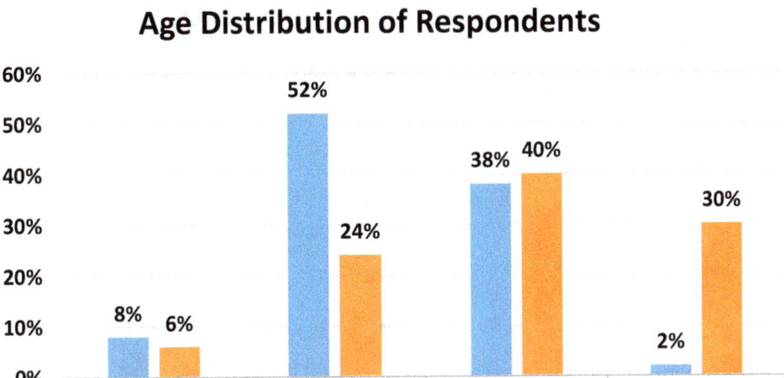

	Below 21	21- 30	31-40	41-50
Owena Osun state	8%	52%	38%	2%
Owena Ondo state	6%	24%	40%	30%

Fig. 4 Age distribution of respondents

Fig. 5 Secondary occupation of respondent

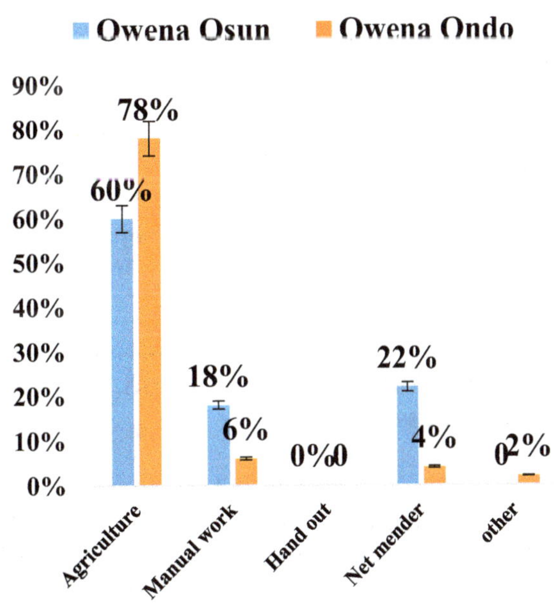

Agriculture was revealed as the major source of income in both States. The chapter revealed that 60% of the respondents in Owena Osun had their secondary occupation as agriculture (fishing), followed by net mending work (22%) and manual work (18%), while in Owena Ondo, agriculture as secondary occupation was 78%, followed by manual work (6%), net mending (4%), and other income sources (2%). The monthly incomes of the respondents from fishing activities as shown in this chapter were as follows: 50% of the respondents in Owena Osun earn

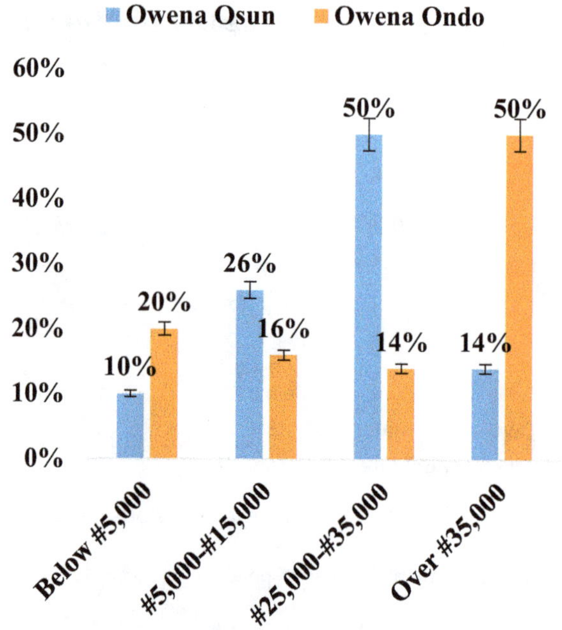

Fig. 6 Income per month of respondent

between #25,000 and 35,000, 26% between #5000 and 15,000, 14% over #35,000, while 10% earn below #5000. Likewise in Owena Ondo, 50% earn over #35,000, 20% below #5000 naira, 16% between #5000 and 15,000, and 14% between 25,000 and 35,000.

The result in this chapter implies that fishing activities in the area might not be sustaining the living standard of the populace and might be the reason why most of the respondents took agriculture as their secondary occupation.

Effect of Climate Variability on Fish Catch in Owena Ondo and Osun States

The effect of wind, rainfall, and temperature on fish catch in Owena Ondo and Osun States is presented in Table 1.

The result in this chapter revealed that variability in wind patterns often affects fish catch in the aquatic ecosystem. As shown in Table 1, 88% and 90% of the respondents observed that strong winds reduce fish catch in Owena Ondo and Osun State, respectively. During moderate wind, 2% and 10% of the respondents reported very low fish catch, 38% and 30% reported high catch, 20% and 10% observe very high catch, while 34% and 50% observe normal fish catch in Owena Ondo and Osun State, respectively. Also, the impact of waves on fish catch by respondents revealed that 52% causes regular and normal catch in the two States, while 36% and 48% cause low catch in Owena Ondo and Osun, respectively. The availability of fish for harvest in various aquatic environments is often affected by variation in wind

Table 1 Result showing the effect of climate variability on fish catch in Owena Ondo and Osun States

Variables	Very low				Low catch				Normal catch				High catch				Very high catch			
States	Ondo		Osun		Ondo		Osun		Ondo		Osun		Ondo		Osun		Ondo		Osun	
	F	%	F	%	F	%	F	%	F	%	F	%	F	%	F	%	F	%	F	%
Strong wind	14	28	20	40	30	50	25	50	6	12	5	10	0	0	0	0	0	0	0	0
Moderate wind	1	2	0	0	3	5	5	10	17	34	25	50	19	38	15	30	10	20	5	10
Waves	3	6	0	0	15	30	24	48	26	52	23	46	4	8	3	6	2	4	0	0
High rainfall	0	0	15	30	2	4	22	44	7	14	8	16	13	26	4	8	23	46	1	2
Moderate rainfall	0	0	2	4	3	6	3	6	11	22	15	30	23	46	25	50	13	26	5	10
Absence of rainfall	1	2	0	0	2	4	9	18	10	20	4	8	5	10	7	14	32	64	30	60
Flooding	1	2	22	44	2	4	16	32	14	28	3	6	14	28	5	10	19	38	5	10
High temperature	1	2	38	76	30	60	6	12	8	16	3	6	3	6	3	6	8	16	0	0
Moderate temperature	0	0	0	0	3	6	4	8	11	12	26	52	20	40	20	40	16	32	0	0
Low temperature	6	12	2	4	21	42	33	66	23	46	10	20	0	0	0	0	0	0	0	0

patterns. This often results in the destruction of fishing infrastructure, submergence/loss of fishing boats with adverse effects on fisher folks, and reduction in the number of catches. Trotman (2009) and Crandall (2009) opined that changes in wind patterns and extreme weather conditions can be damaging to fish production, distribution, abundance, and the fisheries industries as they cause destruction of the fishing gear, vessels, and biodiversity, thus becoming a major setback to the fishing community. The findings support the report of Jallow et al. (1999) who observed changes in wind pattern and adduced it to changes in climatic regimes.

There was marked variation in the rainfall pattern across the two states with different impacts on fish catch. In Owena Ondo, 46% of the respondents observed very high catch, 26% high catch, and 18% low catch during high rainfall. During moderate rainfall, 46% of the respondents observed high catch, 26% very high catch, 22% normal catch, and 6% low catch. During flooding, 66% of the respondents observed that catch is high, while 6% observe low catch. In Owena Osun, 2% of the respondents observed very high catch, 8% high catch, and 54% low catch during high rainfall. During moderate rainfall, 50% of the respondents observed high catch, 10% very high catch, 30% normal catch, and 10% low catch. 76% of the respondents observed low catch during flooding, while 20% observed high catch. Although changes in rainfall pattern have often proof to have different impacts on the duration of fishing, the high catch observed during high rainfall could be as a result of influx of fresh water which could have led to increase in fish growth and performances. In addition, many aquatic species tend to reproduce and migrate during raining seasons due to water and nutrients availability, thus supporting high biomass and catch.

Most of the respondents in both locations reported low fish catch during high and low temperature. 40% and 88% reported low fish catch during high temperature, while 19% and 6% observed high catch in Owena Ondo and Osun, respectively. Also, during low temperatures, 54% 70% of the respondents reported low fish catch in Owena, Ondo and Osun, respectively. The low fish catch during high and low temperature implies a warming trend in the areas. Changes in temperature can change the dynamics of aquatic environments, thus resulting in changes in the migration patterns of fish and reduced fish landings (African-Action 2007).

Likely Effect of Climate Variability on Fisher Folks in Owena Ondo and Osun States

The likely effect of climate variability such as wind, rainfall, and temperature on fisher folks in both locations is presented in Table 2.

In both locations, most of the respondents (94% and 100% in Owena Ondo and Osun, respectively) observed and stated the adverse effect of strong winds on fisher folks. This impact varies from loss of fishing gear (traps and nets), sickness (body injuries), and death. Majority of the respondents (100% and 98% in Owena Ondo and Osun, respectively) preferred moderate, calm wind for their fishing activities. It was also observed by most of the respondents in both states that wave and stormy conditions lead to loss of fishing equipment.

Table 2 Result showing the effect of climate variability on fisher folks in Owena Ondo and Osun States

Variables	Might lead to fisher folks death				Sickness				Loss of fishing gears				No effect			
States	Ondo		Osun		Ondo		Osun		Ondo		Osun		Ondo		Osun	
	F	%	F	%	F	%	F	%	F	%	F	%	F	%	F	%
Strong wind	0	0	2	4	7	14	0	0	43	86	48	96	3	6	0	0
Moderate wind	0	0	0	0	0	0	0	0	0	0	1	2	50	100	49	98
Waves	1	2	0	0	0	0	0	0	30	60	34	68	19	38	16	32
High rainfall	10	2	16	32	45	90	10	20	14	28	36	72	3	6	2	4
Moderate rain	0	0	0	0	0	0	0	0	0	0	1	2	50	100	49	98
Absence of rainfall	0	0	0	0	0	0	0	0	0	0	0	0	50	100	50	100
Flooding	1	2	22	40	31	62	15	30	13	26	39	78	9	18	2	4
High temperature	1	2	6	12	0	0	1	2	0	0	1	2	50	100	42	84
Moderate temperature	0	0	0	0	0	0	0	0	0	0	1	2	50	100	49	98
Low temperature	1	2	0	0	4	8	2	4	0	0	0	0	46	92	48	96

The results in this chapter also revealed the effect of changes in rainfall pattern in both locations. In Owena Ondo State, 90% of the respondents observed high rainfall which results into sickness such as cold and malaria from mosquito bite on the fisher folks, 28% observed the loss of fishing gear in waters, while 20% revealed fisher folks might die as a result of high rainfall. In Owena Osun, the majority of the respondents (72%) observed the loss of gear during high rainfall, 32% observed that it might lead to the death of fisher folks, while 20% observed various sickness on the fisher folks. The major impact of flooding in both locations was sickness and loss of fishing gears and death.

The adverse effect of climate variability on the fisher folks was clearly evident in this chapter. The impacts of strong wind, waves, rainfall, flooding, and temperature were strongly observed by the respondents to cause sicknesses (malaria, cholera), physical injuries, destruction/loss of fishing gears, and death from boat submergence and climatic shocks. Badjeck et al. (2009, 2010) observed the same effects in their studies and noted it to be inimical to health and safety of the fisher folks and deleterious to sustainable livelihood.

Effect of Climate Variability on Fishing Techniques in Owena Ondo and Osun States

The effect of climate variability on fishing techniques in both locations is presented in Table 3.

This chapter revealed that most of the respondents observed the adverse effect of climate variability on the fishing techniques in both locations. During strong winds and storms, most of the respondents (59.38% and 83.33%) do fishing using other

Table 3 Percentage of respondents on effect of climate variability on fishing techniques in Owena Ondo and Osun States

States	Ondo				Osun			
Variables	Normal fishing method (active)		Other fishing method (passive)		Normal fishing method (active)		Other fishing method (passive)	
	F	%	F	%	F	%	F	%
Strong wind	13	40.63	19	59.38	7	16.67	35	83.33
Moderate wind	19	59.38	11	34.38	34	80.95	7	16.67
Waves	15	46.88	17	53.13	19	45.24	23	54.76
High rainfall	11	34.38	19	59.38	16	38.09	26	61.9
Moderate rain	18	56.25	14	43.75	37	88.1	5	11.9
Absence of rainfall	14	43.75	18	56.25	30	71.43	12	28.57
Flooding	14	43.75	31	62	10	23.81	32	76.19
High temperature	15	46.88	17	53.13	37	88.1	5	11.9
Moderate temperature	22	68.75	10	31.25	39	92.86	3	7.14
Low temperature	19	59.38	13	40.63	32	76.19	10	23.81

fishing method (passive method), while 40.63% and 16.67% use normal fishing method techniques (active method) in Owena Ondo and Osun, respectively. During waves, 53.13% and 54.76% use other fishing method, while 46.88% and 45.24% use normal fishing method in Owena Ondo and Osun, respectively.

The result also revealed that fishing techniques change during high rainfall as 59.38% and 61.90% use the other techniques as opposed to the usual fishing methods in Owena Ondo and Osun, respectively. During flooding, 62% and 76.19% use unusual fishing method, while 43.75% and 23.81% use normal fishing method in Owena Ondo and Osun, respectively.

Most of the respondents (53.13% and 88.10%) reported the use of other methods of fishing during high temperatures, while 46.88% and 11.09% reported the use of the usual methods of fishing in Owena Ondo and Osun, respectively.

The effect of climate variability was visible on the various methods and techniques of fishing. The techniques of fishing referred to as the normal methods of fishing include the use of cast nets, gill nets, hand lines, and beach seine nets that is the active fishing, while the other methods of fishing include trapping, long line fishing, and use of baskets, i.e., the passive fishing.

Effect of Variability in Climate on Choice of Fishing Equipment in Owena Ondo and Osun States

The effect of climate variability on the choice of fishing equipment in both locations is presented in Table 4.

This chapter revealed that strong winds destroy 88% and 66% fishing equipment, 6% and 40% causes sinking of gears in Owena Ondo and Osun state, respectively. 66% and

Table 4 Percentage response of respondents on the effect of climate variability on fishing equipment in Owena Ondo and Osun

Variables	Destruction of fishing gears				Sinking of fishing gears				No effect on equipment			
States	Ondo		Osun		Ondo		Osun		Ondo		Osun	
	F	%	F	%	F	%	F	%	F	%	F	%
Strong wind	44	88	28	56	3	6	20	40	3	6	2	4
Moderate wind	0	0	0	0	0	0	1	2	50	100	49	98
Waves	33	66	20	40	3	6	15	30	16	32	15	30
High rainfall	5	10	15	30	37	74	36	72	12	24	5	10
Moderate rain	0	0	0	0	0	0	1	2	50	100	49	98
Absence of rainfall	0	0	0	0	0	0	2	4	50	100	48	96
Flooding	18	36	30	60	26	52	34	68	15	30	5	10
High temperature	0	0	4	8	0	0	0	0	50	100	46	92
Moderate temperature	0	0	0	0	0	0	0	0	50	100	50	100
Low temperature	0	0	0	0	0	0	0	0	50	100	50	100

40% of the respondents revealed the impact of waves on the destruction of fishing equipment in Owena Ondo and Osun, respectively. Also, 74% and 52% of the respondents showed that fishing equipment sink during high rainfall and high flooding, respectively, in Owena Ondo, while 72% and 68% reported the same scenario in Owena Osun. The effect of climate variability on the fishing equipment was strongly observed in the areas. The loss of physical capital equipment (gears and nets) compounded with a deteriorating financial asset base from strong winds, waves, and high rainfall would have had a significant effect on livelihoods in the areas.

The periods of moderate winds and rainfall were observed by the respondents to be the most convenient and secure for fishing equipment. However, temperature variations were observed to have no effect on the fishing equipment in both locations.

Climatic Trends in Owena River Between 2008 and 2017: Wind Trends and Rainfall Pattern in Owena Osun State Between 2008 and 2017

The statistical records on wind and rainfall data were obtained from the Department of Meteorological/Climate Science Technology Department of the Federal University of Technology, Akure, Nigeria, as presented in this chapter. Figure 7 shows the monthly trend of wind data for Owena Osun (2008–2017); Fig. 8 shows annual rainfall pattern/data for Owena Osun State (2008–2017); and Fig. 9 shows

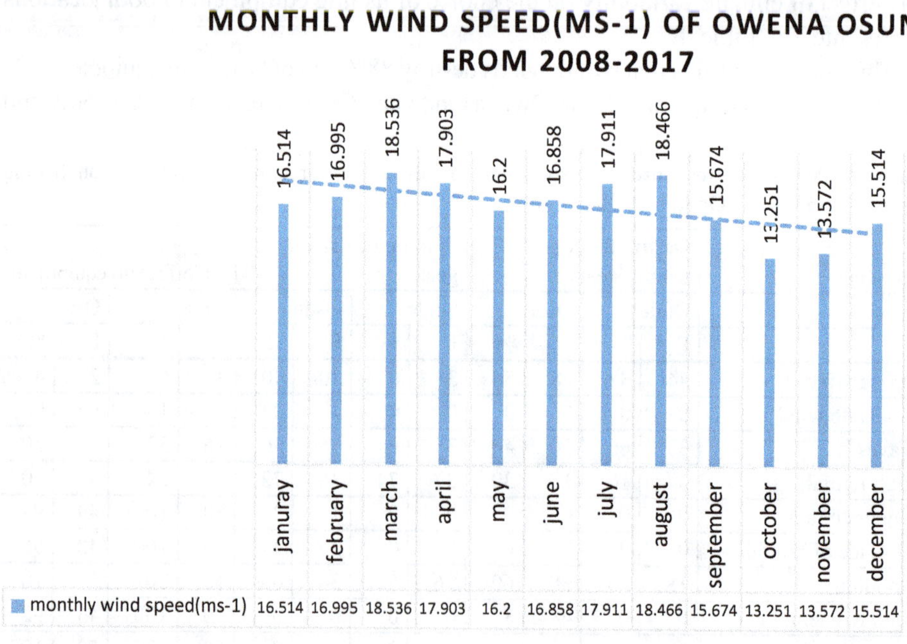

Fig. 7 Monthly trend of wind data for Owena: 2008–2017. (Source: Climate Records of Department of Meteorological/Climate Science Technology, The Federal University of Technology, Akure. Nigeria, 2018)

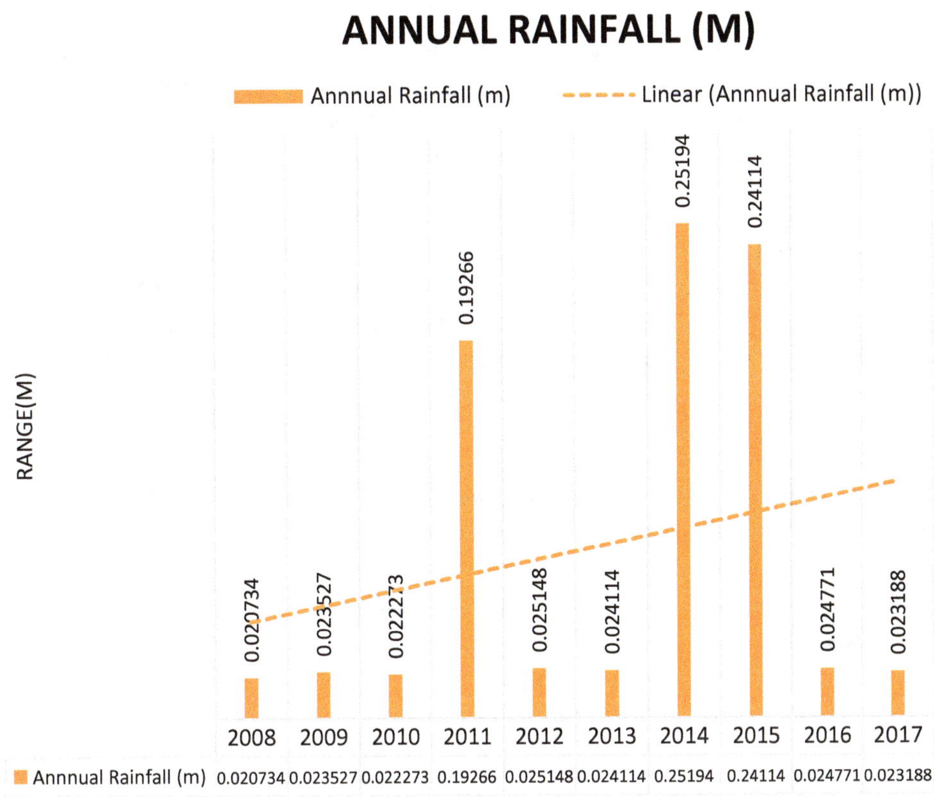

Fig. 8 Annual rainfall pattern/data for Owena Osun State: 2008–2017. (Source: Climate Records of Department of Meteorological/Climate Science Technology, The Federal University of Technology, Akure. Nigeria, 2018)

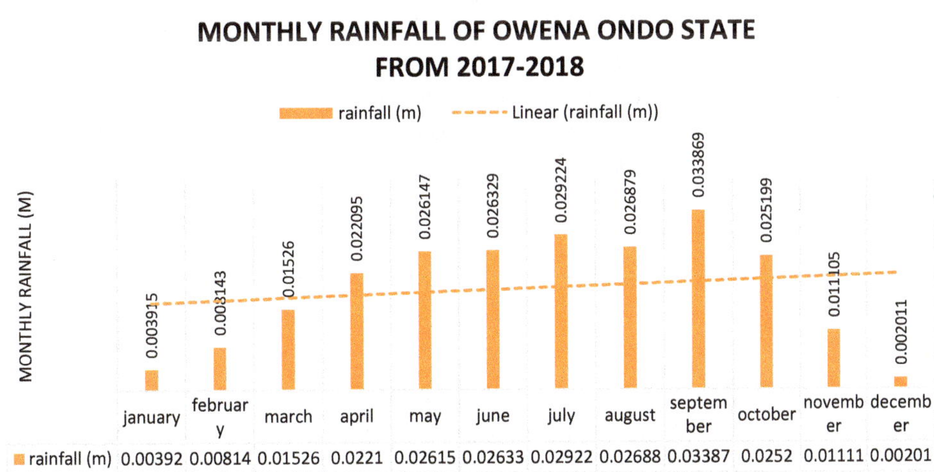

Fig. 9 Monthly rainfall pattern/data for Owena Osun State: 2008–2017. (Source: Climate Records of Department of Meteorological/Climate Science Technology, The Federal University of Technology, Akure. Nigeria, 2018)

monthly rainfall pattern/data for Owena Osun State: 2008–2017. The record reveals over the 10 years duration of the wind trends and rainfall pattern. For the 10 years under consideration (between 2008 and 2017), a decreased trend in the average wind speed across the months from September to December was observed. The highest wind speed was in the month of March and the lowest wind speed in the month of June.

Also, the recorded data on rainfall from 2008 to 2017 showed an increasing trend in annual rainfall. The irregularity in rainfall pattern from these records points a clear indication to weather and climate variability; some years like 2014, 2015, and 2011 have exceptionally high average rainfall. These trends certainly will affect fishing activities. More so, fluctuation in the pattern of rainfall will have both positive and negative impacts on fish catch in the inland Owena water. As opined by Adeleke and Fagbenro (2013), fluctuations in the climate variables will pose adverse effect on the fisher folks and fisheries activities in the coastal areas of Nigeria. In addition, strong winds according to Jallow et al. (1999) have adverse effects on fishing infrastructure as fishing vessels get destroyed, lost, or submerge

Conclusion

This chapter revealed that climate variability has affected fishing activities, fisheries resources, and structures in Owena Osun and Ondo State. The potential impacts of climate change range from impact on the fisher folk, some of which are loss and destruction of fishing equipment's, impact on health, alteration in fishing duration, alteration in fish catch, and change in fishing techniques and equipment. It was also observed that the success and optimum benefit derived from fishing is dependent on favorable climatic conditions. Fisher folk and fishermen that depend on fisheries resources as a source of livelihood will continue to be at risk if proper preventive strategies and measures against the adverse effect of climate and food insecurity are not set in place. This chapter, therefore, recommends the need for extension education and innovation on the various possible strategies, fishing techniques, equipment, and beneficial knowledge to help fisher folks to be able to counter or lessen the adverse effect of climate variability in order for fisher folks to attain maximum sustainable catch.

References

Adeleke ML, Balogun AA (2013) Occurrence of climate change and its effects on the fishing activities in the coastal region of Ondo State, Nigeria. Int J Dev Econ Sustain 1(1):1–13

Adeleke ML, Fagbenro OA (2013) Livelihood diversification and operational techniques of the Artisanal Fisherfolks in the coastal region of Ondo State, Nigeria. Int J Innov Res Dev 2(1):262–273. ISSN 2278-0211(online)

Adeleke ML, Omoboyeje VO (2016) Effects of climate change on aquaculture production and management in Akure Metropolis, Ondo State, Nigeria. Niger J Fish Aquac 4(1):50–58

Africa-Action (2007) Africa policy outlook. Available at http://www.africaaction.org

Agbebi FO, Omoniyi AO (2011) Effect of climate change and coping strategies on socio economic status of fisher folks in Ilaje-Ese Odo local government areas of Ondo State, Nigeria. Int J Sci Nat 2(2):339–343

Allison EH, Perry AL, Badjeck M-C, Adger WN, Brown K, Conway D (2009) Climate change and fisheries: a comparative analysis of the relative vulnerability of 132 countries. Fish Fish 10 (2):173–196

Badjeck M-C, Mendo J, Wolff M, Lange H (2009) Climate variability and the Peruvian scallop fishery: the role of formal institutions in resilience building. Clim Chang 94(1–2):211–232

Badjeck M-C, Allison EH, Halls AS, Dulvy NK (2010) Impacts of climate variability and change on fishery-based livelihoods. Mar Policy 34:375–383

Cheung WWL, Lam VWY, Sarmiento JL, Kearney K, Watson R, Pauly D (2009) Projecting global marine biodiversity impacts under climate change scenario. Fish Fish 10:235–251

Crandall A (2009) Local perceptions of water variability and wetlands in Lake Victoria. Institute for the Study of Diplomacy Edmund A. Walsh School of Foreign Service Georgetown University. Washington DC, USA

Idowu AA, Ayoola SO, Opele AI, Ikenweiwe NB (2011) Impact of climate change in Nigeria. Iran J Energy Environ 2(2):145–152

Jallow BP, Toure S, Barrow MMK, Matheiu AA (1999) Coastal zone of the Gambia and the Abuja Region in cote d'Ivore: sea rise and vulnerability, response strategies, and adaptation options. Clim Res 12(2–3):129–138

Perry RI, Ommer RE, Allison E, Badjeck MC, Barange M, Hamilton L (2009) The human dimensions of marine ecosystem change: interactions between changes in marine ecosystems and human communities. In: Barange M, Field C, Harris R, Hofmann E, Perry I, Werner C (eds) Global change and Marine ecosystems. Oxford University Press, Oxford

Trotman A, Gordon RM, Hutchinson SD, Singh R, McRae-smith D (2009) Policy responses to GEC impacts on food availability and affordability in the Caribbean community. Environ Sci Policy 12(4):529–541

World Fish Center (WFC) (2006) The threat to fisheries and aquaculture from climate change. World Fish Center policy brief. WFC, Penang, p 8

12

Plastic Pollution and Climate Change: Role of Bioremediation as a Tool to Achieving Sustainability

S. A. Idowu, D. J. Arotupin and S. O. Oladejo

Contents

Abstract

Pollution from post-consumer plastics is a growing global environmental challenge whose negative impacts are exacerbating climate change. Plastics are stable, durable, and hydrophobic. They possess high molecular weight, complex three-dimensional structure, and are not readily available to be used as substrate by biological agents such as microorganisms and enzymes. Polyethylene terephthalate (PET) is one of the examples of petrochemical-based plastics. PET is a strong, clear, and light-weight plastic with global usage in the production of bottles.

S. A. Idowu (✉) · D. J. Arotupin
Department of Microbiology, Federal University of Technology, Akure, Nigeria

S. O. Oladejo
Department of Remote Sensing and Geoscience Information System, Federal University of Technology, Akure, Nigeria

Technological innovation, policy formulation, advocacy and sensitization, change in consumption pattern, and bioremediation are some of the approaches that are currently being used to mitigate environmental pollution from post-consumer PET bottles. The ubiquitous property of microorganisms and their ability to survive in almost every environment, including very extreme ones, make them good candidate for biodegradation. Bioremediation is simply defined as engineered or enhanced biodegradation. This review discusses the potential of bioremediation as sustainable and environment-friendly tool to clean up post-consumer PET bottles that already accumulate on land, in soil, and in water bodies.

Keywords

Bioremediation · Plastic · Pollution · Climate change · Biodegradation

Introduction

Plastic is a group of substances with wide global usages and applications as a consumer product. Plastic possesses characteristic cheap price, light weight, and disposable nature which make them the substance of choice in the manufacturing industry. However, plastic pollution is a growing global environmental challenge to aquatic, celestial, and terrestrial organisms including man. The current increasing global human population and technological developments are among the major factors contributing to increasing significant snowballing quantity of plastic waste that is generated annually. This increase exacerbates environmental pollution due to unsustainable methods of disposal of post-consumer (or used) plastics. Accumulation of post-consumer plastics on the surface of the earth including those that are buried in the earth is a serious threat to environmental safety and human health. In general, plastics possess characteristic stable, durable, and hydrophobic nature. They possess high molecular weight, complex three-dimensional structure, and are not readily available to be used as a substrate by many biological agents such as microorganisms and enzymes (Arutchelvi et al. 2008; Kale et al. 2015). As a result of the above, plastics accumulate over a very long period of time. (Kenny et al. 2008) reported that an efficient decomposition of plastic takes about 1,000 years.

Polyethylene Terephthalate (PET): A Typical Plastic

Polyethylene terephthalate (PET) is one of the examples of petrochemical-based plastics. PET is a thermoplastic polyester (Hosseini et al. 2005). It is a strong, clear, and light weight plastic with global usage in the production of bottles. The PET bottles are used for packaging drinks, other food products, and pharmaceuticals. In the field of engineering, PET is being used as alternative or replacement for metals like aluminum, steel, and other metals in the manufacture of precision moldings for office appliances, domestic appliances, and electrical and electronic devices (Hosseini et al. 2005; Tanasupawat et al. 2016; Yoshida et al. 2016). Polyethylene

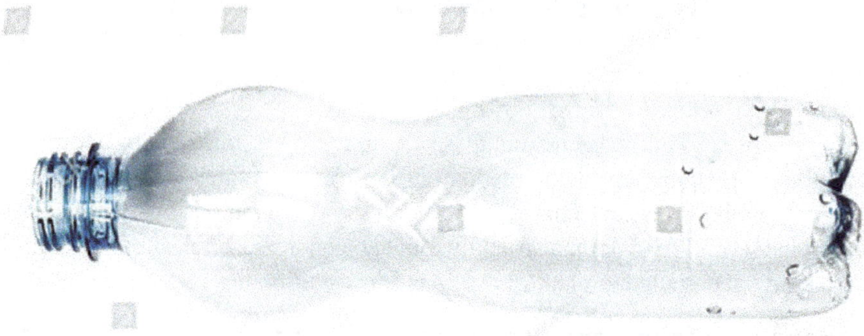

Fig. 1 Typical example of a bottle made from polyethylene terephthalate (PET). (Source: https://www.shutterstock.com/image-photo/top-view-empty-plastic-bottle-isolated-572688928)

terephthalate is a polyester (polymer) produced by polymerization reaction. The reaction involves two monomers, namely terephthalic acid (TPA) and ethylene glycol (EG) according to (Kale et al. 2015). A typical example of bottle made of PET is presented in Fig. 1 below.

The presence of aromatic groups in the PET molecule makes PET nondegradable under normal condition according to the report of (Hosseini et al. 2005). The biological degradation of PET was thought to be limited to only a few fungal species and as a result biodegradation was earlier considered as not yet a remediation strategy for PET (Tanasupawat et al. 2016).

Plastic Waste and Environmental Pollution

The traditional methods (such as burning in the open field) used to dispose of post-consumer plastics, especially in developing countries, is a global environmental concern. Smoke, containing dioxins, furans, mercury, and polychlorinated biphenyls which are injurious to health and the environment, is released into the atmosphere during open burning of plastics, thereby leading to air pollution which consequently contributes to climate change (Webb et al. 2013). In addition, most post-consumer PET bottles end up in landfills where they persist and occupy huge land space, thereby impairing soil fertility with subsequent negative impact on agricultural practice (Chan 2016; Kale et al. 2015). Some post-consumer PET bottles are burnt in open fire during which they release carbon dioxide (CO_2) and dioxins into the atmosphere, also exacerbating environmental pollution and climate change (Kenny et al. 2008; Tanasupawat et al. 2016). Increasing efforts are being made globally to address environmental pollution from PET wastes. Other toxic and hazardous pollutants from plastic wastes are highlighted in the report of (Webb et al. 2013).

Measures and approaches that are currently being used to mitigate plastic pollution include the following:

- Technological innovations (e.g., bioplastics)
- Policies (e.g., ban of single-use plastics and green shopping)
- Change in consumption pattern and different "Rs" including *reduction, reuse,* and *recycling* of post-consumer plastics
- Advocacy/sensitization (e.g., The 2018 World Environment Day with the theme "*Plastic Pollution*").
- Bioremediation (Gnanavel et al. 2012; Hadad et al. 2005; Hosseini et al. 2005; Ron and Eugene 2014; Sardrood et al. 2013; Yoshida et al. 2016)

It is practically impossible to totally eradicate the use of plastics for domestic and industrial uses, neither is it possible not to generate post-consumer plastics due to vast properties of plastics that make them suitable for various domestic and industrial uses. This therefore means that wastes from post-consumer plastics will be generated continually as long as man exists on earth. This waste is projected to increase with the increasing global human population.

Several approaches have been identified to mitigate environmental pollution from post-consumer plastics. However, two of the approaches are majorly relevant to this discourse. The first is a total removal of PET bottles which constitute environmental nuisance. This can be achieved through sustainable clearing of PET bottles that have been used and dumped, mostly in indiscriminate manner, from the surface of the earth or from water bodies. The second approach is replacing plastics of non-biological origin with plastics that are made from biological agents. The latter are commonly referred to as bioplastics. Bioplastics are produced or synthesized from biomass or renewable resources according to the report of (Sardrood et al. 2013). They have been identified to be more readily available for bioremediation than nonbioplastics, thereby preventing accumulation on the environment for a very long period of time. Technology is involved in the production of bioplastics to harness the biomass and other raw materials that are employed in the production. This special type of technological process that involves biological agent is called biotechnology (Sardrood et al. 2013). Detailed discussion of types of biotechnology and the processes involved in each of the types are beyond the scope of this chapter.

Technological advancements and scientific researches toward development and improvement on bioplastics are growing globally. This biotechnology will help to produce plastics that will be readily available for easy degradation by microorganisms, thereby proving solution to environmental and climatic problems that are associated with accumulation of post-consumer plastics. As this biotechnology develops to gain worldwide popularity and acceptance to displace nonbioplastics in circulation, it is important to provide suitable sustainable approach to get rid of post-consumer plastics, particularly PET bottles that are already exacerbating environmental pollution and climate change. A biological approach will be appropriate for this remediation as it is usually environment friendly (Gnanavel et al. 2012; Hosseini et al. 2005). The interesting ability of microorganisms to survive in almost every environment including very extreme ones together with their ubiquitous property makes them a good candidate for bioremediation. Bioremediation is a form of biotechnology which involves engineering. This essence of engineering the process is to enhance the capacity of the microorganisms being used and to make the process faster (Borasiya and Shah 2007; Hadad et al. 2005).

Climate Change

Climate change (natural and human induced) is real. Industrialization and human desire for development and better living, which are carried out sometime in an unsustainable way, affect the environment and natural resources negatively, thereby exacerbating climate change. The realization of this challenge by the General Assembly of the United Nations gave birth to establishment of Brundtland Commission. The Commission in their report tagged *Our Common Future* and submitted to the UN General Assembly in 1987, and identified that both environment and development are inseparable, nevertheless sustainable development must be encouraged in every human development as a means of saving the earth. This development, described popularly as sustainable development, embraces and encourages development that equally promotes economy, social, and environmental developments without prejudice to any of the three members (Burton 1987). Any development whose economic, environmental, and social benefits are not equally addressed is unsustainable and this was part of the recommendation of the commission. Unsustainable consumption of PET bottles is a potential economic, social, and environmental problem that can increase climate change.

Bioremediation

According to (Baggot 1993; Boopathy 2000) in (Sardrood et al. 2013), bioremediation is a process of using living organisms or biological processes to clean up contaminated environments by exploiting and harnessing metabolic abilities of microorganisms to convert contaminants into harmless products by mineralization, generation of carbon (IV) oxide and water, or by conversion into microbial biomass. Bioremediation as evolving environmental biotechnology uses microorganisms in the degradation process and can be optimized to achieve better result (Borasiya and Shah 2007). Some of the numerous applications of bioremediation include cleanup of ground water, sludges, lagoons, and process-waste stream (Boopathy 2000). Practically, bioremediation has been used on a large-scale application in cleanup of oil spill from Exxon in Prince William Sound, Alaska (Kenny et al. 2008). Some of the advantages of bioremediation have been reported (Boopathy 2000; Sardrood et al. 2013).

Microorganisms have the ability to break the molecular chains in polymers like the PET through degradation. Degradation is one of the many processes involved in bioremediation (Sardrood et al. 2013), this means that a biodegradation activity may not result into total bioremediation. The break in molecular chains during biodegradation leads to decrease in the total length of macromolecules that make the polymer and the degree of polymerization (Fig. 3) as contained in the report of (Hosseini et al. 2005).

In Situ and Ex Situ Bioremediation

One of the advantages of bioremediation in management and treatment of wastes is that the process can be carried out at site of contamination or somewhere away from site of contamination. Based on the above, bioremediation can be classified as in situ bioremediation and ex situ bioremediation. The former involves treatment of the contaminated material within the site where contamination or pollution has occurred, while ex situ technique of bioremediation involves physical removal of the contaminant or pollutant from the site of pollution for treatment (Boopathy 2000). In situ bioremediation is usually used in cleanup of oil spills. However, in treating pollutants such as post-consumer PET bottles, ex situ bioremediation may be used.

Organisms Involved in Bioremediation

Bioremediation involves the use of microorganisms in the remediation process. Pollution from post-consumer plastics can be noticed in the air, on land, and in water bodies such as sea and oceans. The natural ability of microorganisms to degrade hydrocarbons, though at a relatively slow rate, makes them suitable candidates for bioremediation. What scientists do is to simply optimize and harness the natural potential of such microorganisms through biotechnology. (Ron and Eugene 2014; Sardrood et al. 2013) highlight how certain species of microorganisms were used to clean oil spills. Several species of fungi, bacteria, and plants are major organisms that have been identified and reported to be involved in bioremediation. There are specific conditions that enhance bioremediation. The conditions must be provided in order for bioremediation process to proceed as planned and expected. According to (Kenny et al. 2008; Ron and Eugene 2014), these favorable conditions include the following:

1. The organism shall be able to live and demonstrate its bioactivity under conditions of pollution
2. A consortium of microorganism that can successfully utilize the pollutant as a substrate must be present
3. Contaminant and the enzymatic system must come in close contact somewhere in or out of the cell
4. The organisms will have the effective enzymes that are important in bioremediation
5. Appropriate favorable environmental conditions must exist or be provided to enhance multiplication of the potential organism to be used for bioremediation

(Boopathy 2000) summarized the key conditions that affect bioremediation as microbial and environmental substrate, aerobic and anaerobic process, growth substrate and co-metabolism physico-chemical bioavailability of pollutants, and mass transfer limitations.

Bioremediation of Polyethylene Terephthalate (PET)

The discovery of the bacterium *Ideonellasakaiensis* 206-F6^T signified another scientific development in bioremediation of PET. The discovery presented *I. sakaiensis* as one of the many microorganisms whose potentials can be harnessed to remediate environmental pollution from post-consumer PET bottles that are disposed indiscriminately. The fungus *Pseudozymajejuensis* isolated from leaves of *Citrus unshiu* in South Korea earlier before the isolation of *I. sakaiensis*, has also been demonstrated to possess remarkable plastic-degrading potential as reported by (Tanasupawat et al. 2016). The fungus possesses the enzyme cutinase which has the ability to degrade some plastics according to (Tanasupawat et al. 2016). Despite the plastic-degrading potential of *P. jejuensis,* the organisms were not reported to degrade PET. This development clearly indicates the need to isolate specific microorganisms that are suitable to degrade each of the large groups of plastics that constitute environmental pollution. This necessitated the need for more scientific researches to isolate microorganisms that posses the ability to utilize PET as their source of carbon. The bacterium, *I. sakaiensis,* is important in bioremediation of post-consumer PET bottles that constitute environmental pollution (Yoshida et al. 2016). According to the report of (Tokiwa et al. 2009; Yoshida et al. 2016), the bacterium showed high PET-degrading potential when compared with previously isolated microorganisms. The authors highlighted factors that favor the bioremediation process in order to achieve optimum result (Fig. 2).

Reports of (Webb et al. 2013; Yoshida et al. 2016) explained that *I. sakaiensis* 206-F6^T possesses critical enzymes that are needed in the degradation of PET. The enzymes include PETase and MHETase. *I. sakaiensis* 206-F6^T uses the two

Fig. 2 Images of plastic pollution on both terrestrial and aquatic environment. (**a**) Dump site showing plastic pollution. (Source: https://www.alamy.com/garbage-dump-plastik-bottles-pet-bottles-image281712766.html). (**b**) Plastic pollution of marine environment. (Source: Saving Earth Encyclopaedia Britannica)

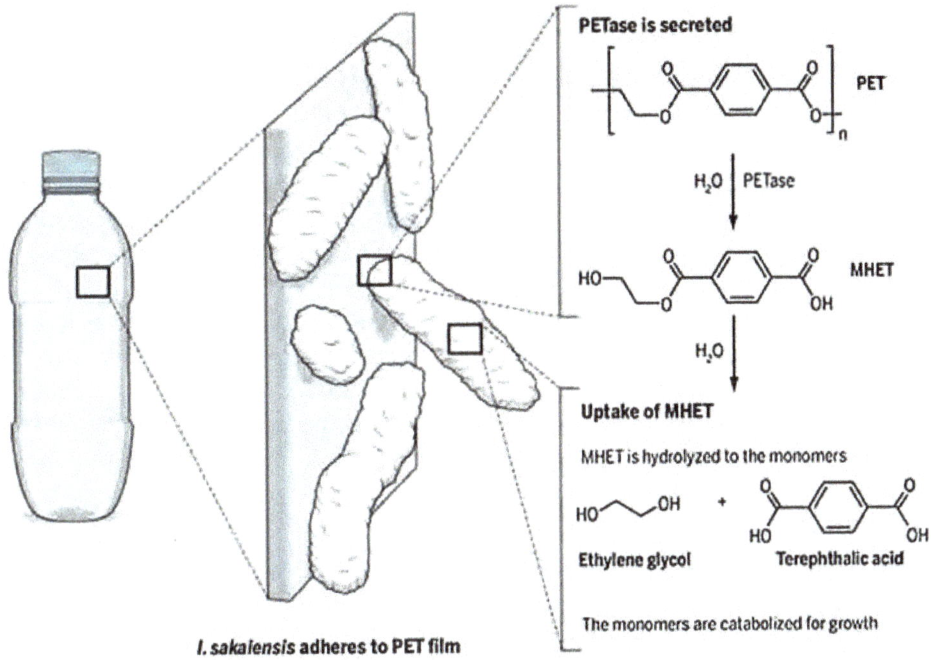

Fig. 3 Schematic degradation of PET bottle by *I.sakaiensis*strain 206-F6^{T}. (Source: (Chan 2016))

enzymes to metabolize PET as its major carbon source (Fig. 3). PETase is a type of esterase and it initiates breaking of the long ester bonds in PET through hydrolytic activity. Esterase is an enzyme that has the ability to break ester bond in a compound. The hydrolytic activity PETase produces an intermediate called mono(2-hydroxyethyl) terephthalic acid (MHET) as presented in Fig. 3 below. The intermediate that results from the hydrolytic breakdown is taken back by the cell of *I. sakaiensis* 206-F6^{T} and is further hydrolyzed by the second enzyme MHETase. The responsibility of MHETase in the degradation process is to break the intermediate into the monomers of PET (Caruso 2015; Tokiwa et al. 2009; Yoshida et al. 2016). The two monomers are named Terephthalic acid (TPA) and ethylene glycol (EG). The report of (Chan 2016) showed that *I. sakaiensis* 206-F6^{T} damaged PET film extensively and almost completely degraded the film after 6 weeks at temperature of 30 °C.

In natural environments, PET bottles, like other plastics, are exposed to UV radiation. The radiation causes cleavage of the C—C bond of the polymer, thereby leading to the formation of low molecular weight fragments. This process increases the susceptibility of the polymer to microbial attack (Arutchelvi et al. 2008). This may be one of the reasons why (Yoshida et al. 2016) used low-crystalline (1.9%) PET film in their study as free-flying plastic bottles in nature may be of a higher crystalline percentage than the ones used by the authors in their research.

Conclusion

The above discourse shows that there are several options to reducing environmental pollution from post-consumer plastics in general and PET bottles in specific. However, bioremediation still stands tall as an important evolving option for sustainable remediation of post-consumer plastics. Other approaches such as discouraging unsustainable consumption of plastics and formulation and implementation of research-informed policies also need to be encouraged as a mean to reducing environmental pollution from plastics.

Microorganisms have the natural potential to use petroleum products including plastics as their sources of carbon and energy, thereby helping to reduce the menace of plastic pollution within the environment and subsequently mitigating climate change. The natural potential of microorganisms can be engineered and optimized through biotechnology as the sustainable way to "clean up" plastic pollution that threatens life on land, water, and air.

It is recommended that in order to further contribute to knowledge on harnessing the potential of microorganisms isolated from PET bottle recycling sites to degrade PET bottles, there is the need to test such microbial isolates with PET film of high crystalline value and be sure that the microorganisms can sufficiently hydrolyze the film. This is to ensure wider global acceptability of the technology because the PET bottles that are used to package food, drinks, and pharmaceutical products are made from high crystalline PET.

References

Arutchelvi J, Sudhakar M, Arkatkar A, Doble M, Bhaddur S, Uppar PV (2008) Biodegradation of polyethylene and polypropylene. Indian J Biotechnol 7:9–22

Baggot J (1993) Biodegradable lubricants. In: Sardrood BP, Goltapeh EM, Varma A (eds) An introduction to bioremediation: fungi as bioremediators. https://doi.org/10.1007/978-3-642-33811-3_1

Boopathy R (2000) Factors limiting bioremediation technologies. Bioresour Technol 74:63–67

Borasiya H, Shah MP (2007) Waste water treatment by environmental microbiology. J Bioremed Biodegr. https://doi.org/10.4172/2155-6199.1000386

Burton I (1987) Report on reports: our common future. Environ Sci Policy Sustain Dev 29(5):25–29. https://doi.org/10.1080/00139157.1987.9928891

Caruso G (2015) Plastic degrading microorganisms as a tool for bioremediation of plastic contamination in aquatic environments. J Pollut Eff Cont. https://doi.org/10.4172/2375-4397.1000e112

Chan A (2016) The future of bacteria cleaning our plastic waste. Berkeley Sci J. https://doi.org/10.5772/48612

Gnanavel G, Valli VPMJ, Thirumarimurugan M (2012) A review of biodegradation of plastics waste. Int J Pharm Chem Sci 1(3):1019–1022

Hadad D, Geresh S, Sivan A (2005) Biodegradation of polyethylene by the thermophilic bacterium brevibacillusborstelensis. J Appl Microbiol 98:1093–1100

Hosseini SS, Taheri S, Zadhoush A, Mehrabani-Zeinabad A (2005) Hydrolytic degradation of poly (ethylene terephthalate). J Appl Polym Sci 102:2304–2309

Kale SK, Deshmukh AG, Dudhare MS, Patil VB (2015) Microbial degradation of plastic: a review. J Biochem Tech 6(1):952–961

Kenny ST, Runic JN, Kaminsky W, Woods T, Babu RP, Keely CM, Blau W, O'Connor KE (2008) Up-cycling of PET (polyethylene terephthalate) to the biodegradable plastic (poly-hydroxyalkanoate). Environ Sci Technol 42(20):7696–7701

Ron E, Eugene R (2014) Enhanced bioremediation of oil spills in the sea. Curr Opin Biotechnol 27:191–194

Sardrood BP, Goltapeh EM, Varma A (2013) In: Goltapeh (ed) Fungi as bioremediators, soil biology. https://doi.org/10.1007/978-3-642-33811-3_1

Tanasupawat S, Takehana T, Yoshida S, Hiraga K, Oda K (2016) Ideonellasakaiensis sp. nov., isolated from a microbial consortium that degrades poly (ethylene terephthalate). Int J Syst Evol Microbiol 66:2813–2818

Tokiwa YA, Calabia BP, Ugwu CU, Aiba S (2009) Biodegradability of plastics. Int J Mol Sci 10:3722–3742

Webb HK, Arnott A, Crawford RJ, Ivanova EP (2013) Plastic degradation and its environmental implications with special reference to poly(ethylene terephthalate). Polymers 5:1–18

Yoshida S, Hiraga K, Takehana T, Taniguchi I, Yamaji H, Maeda Y, Toyohara K, Miyamoto K, Kimura Y, Oda K (2016) A bacterium that degrades and assimilates poly(ethylene terephthalate). Science 351(6278):1196–1199

PERMISSIONS

All chapters in this book were first published by Springer; hereby published with permission under the Creative Commons Attribution License or equivalent. Every chapter published in this book has been scrutinized by our experts. Their significance has been extensively debated. The topics covered herein carry significant findings which will fuel the growth of the discipline. They may even be implemented as practical applications or may be referred to as a beginning point for another development.

The contributors of this book come from diverse backgrounds, making this book a truly international effort. This book will bring forth new frontiers with its revolutionizing research information and detailed analysis of the nascent developments around the world.

We would like to thank all the contributing authors for lending their expertise to make the book truly unique. They have played a crucial role in the development of this book. Without their invaluable contributions this book wouldn't have been possible. They have made vital efforts to compile up to date information on the varied aspects of this subject to make this book a valuable addition to the collection of many professionals and students.

This book was conceptualized with the vision of imparting up-to-date information and advanced data in this field. To ensure the same, a matchless editorial board was set up. Every individual on the board went through rigorous rounds of assessment to prove their worth. After which they invested a large part of their time researching and compiling the most relevant data for our readers.

The editorial board has been involved in producing this book since its inception. They have spent rigorous hours researching and exploring the diverse topics which have resulted in the successful publishing of this book. They have passed on their knowledge of decades through this book. To expedite this challenging task, the publisher supported the team at every step. A small team of assistant editors was also appointed to further simplify the editing procedure and attain best results for the readers.

Apart from the editorial board, the designing team has also invested a significant amount of their time in understanding the subject and creating the most relevant covers. They scrutinized every image to scout for the most suitable representation of the subject and create an appropriate cover for the book.

The publishing team has been an ardent support to the editorial, designing and production team. Their endless efforts to recruit the best for this project, has resulted in the accomplishment of this book. They are a veteran in the field of academics and their pool of knowledge is as vast as their experience in printing. Their expertise and guidance has proved useful at every step. Their uncompromising quality standards have made this book an exceptional effort. Their encouragement from time to time has been an inspiration for everyone.

The publisher and the editorial board hope that this book will prove to be a valuable piece of knowledge for researchers, students, practitioners and scholars across the globe.

LIST OF CONTRIBUTORS

Caroline Fadeke Ajilogba
Division of Agrometeorology, Agricultural Research Council - Soil, Climate and Water, Pretoria, South Africa

Sue Walker
Department of Sociology and Anthropology, Faculty of Social and Management Sciences, University of Buea, Buea, Cameroon
Department of Soil, Crop and Climate Sciences, University of the Free State, Bloemfontein, South Africa

Anthony Njuguna Matheri, Belaid Mohamed
Department of Chemical Engineering, University of Johannesburg, Johannesburg, South Africa

Jane Catherine Ngila
Department of Chemical Science, University of Johannesburg, Johannesburg, South Africa
Academic Affair, Riara University, Nairobi, Kenya

C. A. O. Akinbami
Institute for Entrepreneurship and Development Studies, Obafemi Awolowo University, Ile Ife, Osun State, Nigeria

Peter Asare-Nuamah
Institute of Governance, Humanities and Social Science, Pan African University, Soa, Cameroon
School of Sustainable Development, University of Environment and Sustainable Development, Somanya, Eastern Region, Ghana

Athanasius Fonteh Amungwa
Department of Sociology and Anthropology, Faculty of Social and Management Sciences, University of Buea, Buea, Cameroon

Kwadwo Owusu and Peter Bilson Obour
Department of Geography and Resource Development, University of Ghana, Legon, Ghana

Georgia de Jong Cleyndert, Rebecca Newman, Aida Cuni-Sanchez and Robert Marchant
York Institute for Tropical Ecosystems, Department of Environment and Geography, University of York, York, North Yorkshire, UK

Cecile Brugere
Soulfish Research and Consultancy, York, UK

Yaya Idrissou, Alassan Seidou Assani and Ibrahim Alkoiret Traoré
Laboratoire d'Ecologie, Santé et Production Animales (LESPA), Faculté d'Agronomie, Université de Parakou, Parakou, République du Bénin

Mohamed Nasser Baco
Laboratoire Société-Environnement (LaSEn), Faculté d'Agronomie, Université de Parakou, Parakou, République du Bénin

Nkiru Theresa Meludu and Toyin Abolade
Department of Agricultural Economics and Extension Faculty of Agriculture, Nnamdi Azikiwe University, Awka, Nigeria

M. L. Adeleke
Department of Fisheries and Aquaculture Technology, The Federal University of Technology, Akure, Akure, Nigeria

D. Al-Kenawy, A. M. Nasr-Allah, M. Dickson
WorldFish, Abbassa, Abou-Hammad, Sharkia, Egypt

Aliyu Barau and Aliyu Sani Wada
Department of Urban and Regional Planning, Bayero University Kano, Kano, Nigeria

Lydia Adeleke
Department of Fisheries and Aquaculture Technology, Federal University of Technology, Akure, Nigeria

Jacob Victor Jerry, Akinola Joshua Oluwatobi and Ayodele Idowu Sunday
Department of Fisheries and Aquaculture Technology, Federal University of Technology, Akure (FUTA), Akure, Nigeria

Desalegn Ayal
Center for Food Security Studies, College of Development Studies, Addis Ababa University, Addis Ababa, Ethiopia

Ajibefun Igbekele Amos
Department of Agricultural and Resource Economics, Federal University of Technology, Akure (FUTA), Akure, Nigeria

S. A. Idowu, D. J. Arotupin
Department of Microbiology, Federal University of Technology, Akure, Nigeria

S. O. Oladejo
Department of Remote Sensing and Geoscience Information System, Federal University of Technology, Akure, Nigeria

Index

Printed in the USA
CPSIA information can be obtained
at www.ICGtesting.com
LVHW082205201023
761596LV00009B/1585

9 781647 403416